00-HHE-5T

Identifying and Controlling Odor in the Municipal Wastewater Environment Phase II: Impacts of In-Plant Parameters on Biosolids Odor Quality

by:

Gregory A. Adams
LACSD (Co-Principal Investigator)

Jay Witherspoon
CH2M HILL (Co-Principal Investigator)

Tom Card
Environmental Management Consulting

Zeynep Erdal
Bob Forbes
David McEwen
CH2M HILL

Jim Geselbracht
Damon S. Williams Associates

Dietmar Glindemann
Virginia Polytechnic Institute & State University

Ron Hargreaves
LACSD

Larry Hentz
PES&J

Matthew Higgins
Bucknell University

Sudhir Murthy
DC-WASA

2003

Water Environment Research Foundation

The Water Environment Research Foundation, a not-for-profit organization, funds and manages water quality research for its subscribers through a diverse public-private partnership between municipal utilities, corporations, academia, industry, and the federal government. WERF subscribers include municipal and regional water and wastewater utilities, industrial corporations, environmental engineering firms, and others that share a commitment to cost-effective water quality solutions. WERF is dedicated to advancing science and technology addressing water quality issues as they impact water resources, the atmosphere, the lands, and quality of life.

For more information, contact:
Water Environment Research Foundation
635 Slaters Lane, Suite 300
Alexandria, VA 22314-1177
Tel: (703) 684-2470
Fax: (703) 299-0742
www.werf.org
werf@werf.org

This report was co-published by the following organizations. For non-subscriber sales information, contact:

Water Environment Federation
601 Wythe Street
Alexandria, VA 22314-1994
Tel: (800) 666-0206
Tel: (703) 684-2452
Fax: (703) 684-2492
www.wef.org
pubs@wef.org

IWA Publishing
Alliance House, 12 Caxton Street
London SW1H 0QS, United Kingdom
Tel: +44 (0) 20 7654 5500
Fax: +44 (0) 20 7654 5555
www.iwapublishing.com
publications@iwap.co.uk

Library of Congress Catalog Card Number: 2003096144

IWAP ISBN: 1-84339-687-4 WEF ISBN: 1-57278-187-4

Organizations that helped prepare this report LACSD, CH2M HILL

The research on which this report is based was funded, in part, by the United States Environmental Protection Agency through Cooperative Agreement No. CR-827345-4 with the Water Environment Research Foundation (WERF). Unless an EPA logo appears on the cover, this report is a publication of WERF, not EPA. Funds awarded under the Cooperative Agreement cited above were not used for editorial services, reproduction, printing, or distribution.

ACKNOWLEDGMENTS

Impacts of In-Plant Parameters on Biosolids Odor Quality, representing Phase II of the WERF project, *Identifying and Controlling Odor in the Municipal Wastewater Environment* (WERF 00-HHE-5), was a significant effort undertaken by a large number of participants, whose names and affiliations are listed below.

In addition to the project team, the following agencies provided their wastewater treatment plants as test sites for the project:

- Hampton Roads Sanitation District, Virginia
- Bureau of Sanitation, Department of Public Works, City of Los Angeles, California
- Los Angeles County Sanitation Districts, California
- Philadelphia Water Department, Pennsylvania
- Public Utilities Commission, City and County of San Francisco, California
- South Bayside Systems Authority, California
- Water Pollution Control Division, City of Toronto, Ontario, Canada
- U.S. Filter Operations Services

We would also like to acknowledge and thank the following members of our Technical Advisory Committee for their review and guidance:

Glen Daigger, P.E., Ph.D., *CH2M HILL*
Lawrence Koe, P.E., Ph.D., *National University of Singapore*
Thomas Mahin, B.S., *Massachussetts DEP*
Charles M. Murray, M.S., *Washington Suburban Sanitary Commission*
Phillip L. Wolstenholme, B.S., *Brown and Caldwell*

Report Preparation

Principal Investigators:

Gregory M. Adams, P.E., M.S., *LACSD (Co-Principal Investigator)*
Jay Witherspoon, P.E., M.S., *CH2M HILL (Co-Principal Investigator)*

Project Team:

Tom Card, P.E., M.S., *Environmental Management Consulting*
Zeynep Erdal, Ph.D., *CH2M HILL*
Bob Forbes, P.E., M.S., *CH2M HILL*
Jim Geselbracht, P.E., M.S., *Damon S. Williams Associates*
Dietmar Glindemann, Ph.D., *Virginia Polytechnic Institute & State University*
Ron Hargreaves, P.E., M.S., *Los Angeles County Sanitation Districts*
Larry Hentz, P.E., M.S., *Post, Buckley, Schuh & Jernigan, Inc.*
Matthew Higgins, Ph.D., *Bucknell University*
David McEwen, P.E., M.S., *CH2M HILL*
Sudhir Murthy, P.E., Ph.D., *District of Columbia Water & Sewer Authority*

Project Subcommittee

Michael Jawson, Ph.D., *USDA – ARS – NPS (Chair)*
Andrew Chang, Ph.D., *University of California, Riverside*
Jane Forste, M.S., *Jane Forste Associates*
Jerry Hatfield, Ph.D., *USDA ARS National Soils Tilth Lab*
Lynn Szabo, B.S., *DuPont Engineering*
John Walker, Ph.D., *U.S. Environmental Protection Agency*

Water Environment Research Foundation Staff

Director of Research: Daniel M. Woltering, Ph.D.

Project Manager: Lola Olabode, M.P.H.

ABSTRACT

This project was undertaken in response to the wastewater treatment industry's need to better understand the generation of odors from biosolids produced by wastewater treatment plants (WWTPs). Its primary objective is to begin to establish relationships between WWTP process parameters and biosolids odors, so that more effective techniques for minimizing biosolids odors can be developed.

The project consisted of a detailed field study involving extensive sampling and analyses at 11 WWTPs across North America with capacities from 13 to 350 million gallons per day. Biosolids samples were collected from the WWTPs at a number of sampling points, which were chosen to represent a complete snapshot of biosolids generation and handling at each facility. The sampling points started with influent wastewater, proceeded through primary and secondary clarification, through digestion, dewatering, and onsite storage of dewatered biosolids cake.

Laboratory-scale anaerobic storage tests were conducted to simulate odor development of biosolids in storage prior to their beneficial reuse or disposal. A battery of analyses was performed on the biosolids samples by the participating utility laboratories, commercial laboratories, and specialized university laboratories. The analytical data were evaluated and compared with process and operation parameters at each participating WWTP.

Results indicate that the anaerobic digestion process, including its impacts on achieving stability and minimizing odors in the final biosolids product, are not yet completely understood. A significant finding was that biosolids odors after digestion and dewatering correlate with the amount of bio-available protein in the biosolids. Possible causes for increased bio-available protein and increased odor generation from dewatered biosolids begin with the primary and secondary sludge handling, mixing, and liquid storage steps, and continue through the anaerobic digestion process to post-digestion processes, such as dewatering, conveyance, and cake storage.

A list of future research needs that was developed based on the study findings centered on the need for more controlled experiments to identify and quantify the impacts of different biosolids handling and stabilization processes on biosolids odor generation.

BENEFITS AND KEYWORDS

Benefits:

- Helps the wastewater treatment industry understand and manage biosolids odor and its impacts on surrounding communities by understanding more completely the chain of events involved in the generation of biosolids odors.
- Identifies gaps in scientific knowledge regarding mechanisms of odor generation in WWTP biosolids.
- Shows that biosolids stability parameters may be misleading with respect to their impact on odors produced from biosolids.
- Provides a reference guide for the wastewater treatment industry and a starting point in identifying the causes of biosolids odors.
- Emphasizes the importance of whole plant management for reduction and control of biosolids odors.

Keywords: anaerobic digestion, biosolids, odors, olfactometry, wastewater, WWTP

TABLE OF CONTENTS

LIST OF TABLES

LIST OF FIGURES

LIST OF ACRONYMS

Al	Aluminum
APHA	American Public Health Agency
ASTM	American Society for Testing and Materials
AVSR	Additional volatile solids reduction
BOD	Biological oxygen demand
BRC	Biosolids Recycling Center
°C	Degrees Celsius
Ca	Calcium
CEN	European Committee for Standardization
CLP	Contract Laboratory Program
CO_2	Carbon dioxide
COS	Carbonylsulfide
CS	Combined sludge
CS_2	Carbondisulfide
DAF	Dissolved air flotation
DAFT	Dissolved air flotation thickener
DMDS	Dimethyldisulfide
DMS	Dimethylsulfide
DQO	Data quality objectives
DS	Dry solids
DSWA	Damon S. Williams and Associates
DT	Detection Threshold
D/T	Dilutions-to-Threshold
EMC	Environmental Management Consulting
Fe	Iron
FID	Flame ionization detector
g	Grams
GBT	Gravity belt thickener
GC	Gas chromatograph
GC/MS	Gas chromatograph/mass spectrophotometer
H_2S	Hydrogen sulfide
HCl	Hydrochloric acid
HP	Hewlett Packard
HPLC	High Performance Liquid Chromatography
HT	Hedonic tone
I	Odor intensity
ICP	Inductively coupled plasma
ID	Internal diameter
K	Potassium
kg	Kilogram
L	Liter
LACSD	Los Angeles County Sanitation Districts

LLAP	l-leucine aminopeptidase
M/D ratio	Monovalent to divalent cation ratio
MBTH	3-methyl-2-benzothiazolinone hydrazone
μm	Micrometers
μg	Micrograms
meq/L	Milliequivalents per liter
METase	L-methionine-v-lyase
MG	Million gallons
mg/g	Milligrams per gram
mg/L	Milligrams per liter
mg/m^3	Milligrams per cubic meter
mgN/m^3	Milligrams of nitrogen per cubic meter
$mg\ S/m^3$	Milligrams of sulfur per cubic meter
mgd	Million gallons per day
mL	Milliliter
mV	Millivolts
MT	Methanethiol or methyl mercaptan
N	Normal
Na	Sodium
NaOH	Sodium hydroxide
NH_3	Ammonia
NH_4	Reduced ammonia
nm	Nanometers
ORP	Oxidation-reduction potential
PBS	Phosphate buffer saline
PBS&J	Post, Buckley, Schuh & Jernigan, Inc.
PET	Polyethylene terephthalate
PETE	Polyethylene terephthalate ester
POTW	Publicly owned treatment work
ppbv	parts per billion by volume
ppmv	Parts per million by volume
ppm	Parts per million
PS	Primary solids
PSC	Project Subcommittee
psi	Pounds per square inch
QA	Quality assurance
QC	Quality control
RAS	Return activated sludge
RBA	Residual biological activity
RC	Recessed chamber
RFI	Request for information
RPD	Relative percent difference
rpm	Revolutions per minute
RT	Recognition Threshold
SOP	Standard operating procedure
SPME	Solid Phase Membrane Extraction

SRT	Solids retention time
TAC	Technical Advisory Committee
TCD	Thermal conductivity detector
TCR	Targeted collaborative research
TCS	Thickened combined sludge
t_d	Detention time
TKN	Total Kjeldahl nitrogen
TMA	Trimethylamine
TRL	Target Reporting Limit
TS	Total solids
TSS	Total suspended solids
TWAS	Thickened waste activated sludge
U.S. EPA	U.S. Environmental Protection Agency
UV	Ultraviolet
V/V	volume per volume
VAR	Vector attraction reduction
VFA	Volatile fatty acid
VPI	Virginia Polytechnic Institute and State University
VS	Volatile solids
VSC	Volatile sulfur compound
WAS	Waste activated sludge
WEFTEC	Water Environment Federation Technical Exposition and Conference
WERF	Water Environment Research Foundation
WWTP	Wastewater treatment plant

EXECUTIVE SUMMARY

ES.1 Introduction

The project summarized in this report, *Impacts of In-Plant Parameters on Biosolids Odor Quality*, represents Phase II of a larger project by the Water Environment Research Foundation (WERF) called *Identifying and Controlling Odor in the Municipal Wastewater Environment* (WERF 00-HHE-5). The project to date has been comprised of two major study phases. Phase I was a review of literature related to odors in the wastewater industry and has been published separately. Phase II was a field and laboratory study of plant parameters related to odors from biosolids and is the subject of this report.

ES.2 Objectives of this Study

In Phase II the project team established as its primary goal determining how process conditions (storage, anaerobic digestion, and mechanical dewatering) affect odor emissions from biosolids in wastewater treatment facilities. In accomplishing this goal, the project team set the following objectives:

- Produce a consistent set of general testing protocols to be followed at identical testing events at every facility in the study.
- Use established and new sampling and analytical methods to measure odor precursors in the liquid and gaseous phases of the biosolids, which were produced under a variety of set process conditions.
- Enter process and operational data from all plants in the study for the week and month prior to the testing date into a Request for Information database.
- Draw correlations between the process conditions and the measured odor precursors to provide a better understanding of the conditions that produce more odorous biosolids.

Phase II of the project was developed to be an *observational study* of biosolids odor characteristics, summarizing detailed field work and laboratory analyses of samples collected from 11 wastewater treatment plants (WWTPs) across North America. The main purpose of the study was to observe and document relationships and correlations found among wastewater characteristics, plant operations, and biosolids odor characteristics.

ES.3 Hypotheses Supported, Rejected, or Found Inconclusive

Various hypotheses concerning the origins of odors in anaerobically digested biosolids have been put forth as a result of prior research (see References). The research findings for this study have been linked to these hypotheses and grouped as being supported with conclusive evidence, rejected with conclusive evidence, or inconclusive, depending on the sufficiency of information from the study.

ES.3.1 Hypotheses Supported Based on Study Results

The list below contains hypotheses derived from previous research and experience that were found to be conclusively supported based on the results and correlations developed as part of the study:

1. Higher amounts of bio-available (labile) protein in biosolids cake create more odors.
2. Different dewatering practices affect bio-available protein differently; some dewatering practices tend to increase odors in the biosolids cake.
3. Volatile sulfur compounds (VSCs) are the major sources of odors in digested biosolids. This relationship was shown by a high correlation between odor detection threshold (DT) and concentration of VSCs, indicated by a multiple regression equation having a correlation coefficient of 0.90, which describes this relationship.
4. Odor concentrations in mesophilicly digested biosolids cake rise and then decline over time during storage.
5. Based on comparison of results from the one WWTP in the study with thermophilic anaerobic digestion and 10 WWTPs with mesophilic digestion, odors from thermophilicly digested biosolids cake have different characteristics and patterns of time release than mesophilicly digested biosolids cake.
6. Iron in sufficient concentrations binds bio-available protein in biosolids cake and thus reduces odor production from dewatered biosolids.

ES.3.2 Hypotheses Rejected Based on Study Results

The list below contains hypotheses that were developed because of a collective belief in the industry that a potential relationship exists, as reported in the literature. However, the results of the study indicated that no relationship exists, and therefore the hypotheses were rejected based on the data collected from 11 WWTPs studied and the correlations that were produced.

1. The study findings showed no positive correlation between high influent sulfate concentrations and odors in biosolids cake.
2. The study findings provided no evidence that WAS has a higher odor potential following digestion than primary sludge.
3. The study findings provided no evidence that enzyme activity can be used as an indicator of biosolids odor production.

ES.3.3 Hypotheses Found to be Inconclusive Based on Study Results

The list below contains hypotheses developed because the project team believed, based on prior research and experience, that potential relationships exist between biosolids processes or characteristics and odors emitted. While some results of the study support these hypotheses, definite conclusions could not be established based on the data compiled and evaluated in the study. Further investigation is needed to either confirm or refute these inconclusive relationships.

1. Past experience has indicated that high-pressure conveyance of biosolids after dewatering leads to increased odors in the biosolids cake. Based on conflicting evidence from this study, this hypothesis could not be supported or rejected.
2. Past experience has indicated that the presence and relative concentration of metal cations in the influent wastewater reduces odors in dewatered biosolids cake. In this study there was little correlation with either influent cation concentration or influent monovalent to multivalent (M/D) ratio and odor reduction in the dewatered biosolids cakes sampled.
3. Past experience has indicated that mixing primary and secondary biosolids in pre-digestion processes worsens odors in digested, dewatered biosolids. Some plants in this study showed this correlation, but others did not; hence this relationship cannot be fully supported.
4. Other experience has indicated that higher concentrations of metal cations such as iron could reduce odors in digested, dewatered biosolids. There appears to be a lack of knowledge about the bioavailability of proteins and other organic material complexed with cations such as iron and aluminum, causing the project team to remain undecided about this hypothesis.
5. Past experience indicates that longer primary sludge detention time prior to digestion may increase odors in digested biosolids. This correlation was not supported by the results of this study.
6. Past experience indicates that a longer WAS solids retention time (SRT) in the treatment process produces less odors in digested, dewatered biosolids. This relationship was not seen in this study; however, all plants tested in this study had a very low WAS SRT. Therefore, more plants with longer WAS SRTs must be tested to confirm or refute this hypothesis.
7. Title 40 of the Code of Federal Regulations Part 503 (503 Rule) for vector attraction reduction requires achieving a minimum of 38% volatile solids destruction during digestion. However, a clear correlation between biosolids stability per this criterion and reduction in odors from dewatered biosolids was not found in this study.
8. Based on the study results it could not be shown that trimethylamines (TMA) are significant sources of odor in anaerobically digested, non-limed biosolids.

ES.4 Background

Phase I of the project consisted of a comprehensive review of published literature and unpublished or gray literature related to the identification and control of odors in the wastewater treatment industry and related industries. The papers that were identified and reviewed dealt with odors generated by a wide range of municipal wastewater treatment plant processes, including collection systems and biosolids processing facilities. Related papers dealing with odor identification and control at industrial facilities and the agricultural industry were also reviewed. In October 2000, a specialty workshop was held at the 2000 Water Environment Federation Technical Exposition and Conference (WEFTEC) to review the Phase I findings and prioritize the types of odor problems that were most important to the workshop attendees.

The highest-ranked issue for study in Phase II was to conduct research that would improve understanding of the impact of in-plant operational practices on biosolids odor quality. The common element in this Phase II study was the anaerobic digestion process, since anaerobic digestion is used to stabilize the largest proportion of biosolids from wastewater treatment plants in the United States. The general belief in the wastewater treatment industry that anaerobic digestion reduces odor emissions by stabilizing the solids was not explicitly substantiated by the information gathered during Phase I.

The project team decided that a full-scale, comparative field and laboratory study would be necessary to assess more accurately the processes and mechanisms responsible for the generation of odors from biosolids, and to provide a better direction for future research in the area of biosolids stability and odors. After the development of testing protocols and a formal solicitation phase, the project team selected eight wastewater utilities to participate in Phase II. Three of the eight organizations offered two WWTPs each for sampling and onsite testing, for a total of 11 participating WWTPs. The eight organizations that provided in-kind services and project funding were:

1. Hampton Roads Sanitation District, Virginia
2. Bureau of Sanitation, Department of Public Works, City of Los Angeles, California
3. Sanitation Districts of Los Angeles County, California
4. Philadelphia Water Department, Pennsylvania
5. Public Utilities Commission, City and County of San Francisco, California
6. South Bayside Systems Authority, California
7. Water Pollution Control Division, City of Toronto, Ontario, Canada
8. U.S. Filter Operating Services/Veolia Water North America

Testing laboratories from Virginia Polytechnic Institute and State University (VPI) and Bucknell Universities performed the chemical analyses of biosolids samples obtained from the participating WWTPs. St. Croix Sensory Laboratories performed the olfactometry analyses of selected samples from the WWTPs.

ES.5 Phase II Protocol Development

The project team developed a general but in-depth testing and sampling protocol to encompass plant operating parameters monitored at sampling points throughout the process. The general protocol for sampling locations was customized into a site-specific protocol for each participating WWTP. Since each plant had a number of unique operational points, the project team modified the general protocol to allow samples to be representative of the processes both upstream and downstream of the anaerobic digestion process.

The sampling team loaded identically weighed samples of liquid and cake biosolids into empty 500 milliliter (mL) polyethylene terepthalate (PET) bottles. The bottles were sealed and incubated for up to 40 days at 22°C for specialized analyses. Approximately 150 sample bottles were prepared for each WWTP, in addition to the samples collected and prepared for analysis by Standard Methods by each WWTPs local laboratory. Table ES-1 summarizes the sample types collected from different points in the treatment process, and the laboratories responsible for each test listed.

Table ES-1. Sampling and Analyses Matrix as a Function of Sample Location.

Analysis	Sample Location									
	Influent	Pre Storage WAS A	Pre Storage Primary B	Post Storage Primary C	Mixed WAS and Primary D	Digester Off Gas E	Digester Effluent F	Dewatered Cake G	Dewatering Underflow H	Stored Cake I
Sample Type	Liquid	Liquid	Liquid	Liquid	Liquid	Gas	Liquid	Cake	Liquid	Cake
Standard Tests (performed by local utilities)[1]										
Solids (VS & TS)	X	X	X	X	X		X	X	X	X
ORP/pH/ temp	X	X	X	X	X		X	X	X	X
Alk/NH_3/ TKN	X	X	X	X	X		X	X	X	X
VFA/ Coliforms							X			X
VS Destr/ Methane					X	X	X			
Field Odor Tests (onsite by WERF Project Team)										
Colorimetry H_2S						X				
Jerome H_2S	X	X	X	X	X		X	X	X	X
Colorimetry NH_3	X	X	X	X	X		X	X	X	X
Headspace Analysis (VPI)[3,4]										
Reduced sulfur	X	X	X	X	X		X	X	X	X
Amines	X	X	X	X	X		X	X	X	X
Ketones	X	X	X	X	X		X	X	X	X
Olfactometry[5]							X	X		X
Organic Compounds Analysis (Bucknell)[2]										
Proteins			S,T	S,T	S,L,T		S,L,T	L,T	S,T	L,T
Enzyme Assays					L		L	L		L
Amino Acids					T		S,L,T	L		L
VFA				S	S		S			L
Cations & Anions (Bucknell)[2]										
Fe/Al	T				T		T		T	
Ca/Mg/Na/K	S	S			S,T		S,L,T	L	S	L
Sulfate			S	S	S					
Sulfide							S,L	L		L
Residual Biological Activity (VPI)[2]										
Additional VS Destruction							X	X		X
Methane							X	X		X
Ammonia: the extraction yields soluble + bound NH_4							T	T		T

Notes:

1. X = Sampling and analysis point at that location
2. Employ analysis of fractions: S = soluble, L = labile, T = total NaOH extract
3. Repeated analysis of samples from Sample Locations F, G, and I, every day for 1 week of lab incubation.
4. Measurement of analytes that require longer gas chromatograph (GC) times were conducted 48 hours after sampling and onwards for samples obtained from Sample Locations F, G, and I only.
5. Analysis by St. Croix Sensory Labs on Day 7 after sampling at VPI on Day 6 of lab incubation.

ES.6 Sampling, Laboratory Analysis, and Data Collection

Sampling events at each of the 11 WWTPs occurred between May 22 and August 23, 2002, while routine laboratory analyses continued for a month after the last sampling event. Some additional laboratory tests were run after an initial comparison of the analytical results in late September and early October. The project team collected, compiled, and analyzed Request for Information (RFI) files from the participating WWTPs. These files contained detailed information regarding design parameters, physical plant characteristics, flows and loadings, and process information calculated using some of the requested data.

The project team evaluated the large amount of data using several statistical analysis techniques and examined numerous relationships between analytical parameters and process data to evaluate potential correlations between plant parameters and biosolids odor generation. Table ES-2 lists pertinent process data from the 11 WWTPs that took part in this study, focusing on biosolids production, handling, and treatment facilities at the WWTPs. More detailed information on each WWTP is included in Chapter 3.0 and the Appendix.

Table ES-3 contains information collected by the project team during the sampling events at each WWTP. The participating WWTPs ranged in size from 13 million gallons per day (mgd) of influent flow to 350 mgd. Field analyses performed included pH, temperature, oxidation reduction potential (ORP), headspace hydrogen sulfide concentration, and headspace ammonia concentration at each sample location. Some of the WWTPs used two-phase digestion or post-digestion biosolids storage. In those cases, "Digester ORP" data presented in Table ES-3 shows two values in one cell, the first value measured in the primary digester effluent and the second in the secondary digesters or post-digestion storage tanks.

Table ES-4 summarizes the data gathered as a result of the analytical work performed at the St. Croix Sensory Laboratories (olfactometry analyses) and VPI's Environmental Engineering Department Analytical Laboratories (headspace gas analyses for odorous compounds quantification). The olfactometry analysis represented odor emitted from samples collected over a six-day accumulation period in the sample headspace bottles. Based on the experience of the project team members, Day 6 was chosen to be representative of a storage period during which most odorous compounds would have been released.

The two fractions of proteins reported in Table ES-5 are "labile" (also known as "bio-available") protein and "sodium hydroxide (NaOH)-extracted" protein. These fractions of protein were measured in the solids fraction of the cake samples rather than in the liquid. Soluble protein fraction was also measured as part of this study, but soluble protein is usually removed during the dewatering process, so it is not reported in Table ES-5. The project team calculated digester volatile solids (VS) destruction values based on WWTP data provided by the participating agencies. The VPI research laboratory analyzed for residual biological activity (RBA) and reported RBA in the form of the test for additional volatile solids reduction (AVSR) described in the 503 Rule as Option No. 2 for meeting the requirements for vector attraction reduction (VAR). This VAR Option No. 2 is a bench-scale test to determine the biological stability of anaerobically digested biosolids. WWTP No. 2 has two values for the dewatered biosolids RBA, which is due to the presence of two types of dewatering centrifuges, as listed in Table ES-2.

Table ES-2. Plant Information[1].

		Plant 1	Plant 2	Plant 3	Plant 4	Plant 5	Plant 6	Plant 7	Plant 8	Plant 9	Plant 10N	Plant 10S	Plant 11
Primary Sludge t_d	hr	214	217	197	198	2	83	1	37	43	31	31	165
Activated Sludge SRT	D	1.7	1.6	0.98	10.2	5.9	2.9	3.89	1	6.4	1.5	1.5	1.9
WAS VS	%	82	71	77	53	76	78	80	80	81	75	75	70
WAS Thickening	Type	DAFT	DAFT	DAFT	Belt	Belt	Gravity Thickener	Gravity Thickener	Centrifuge	DAFT	DAFT	DAFT	DAFT
WAS Thickening t_d	Hr	2.5	7.5	7.7	0	0	14	15.2	0	2.5	0.6	-	1.3
Digester Feed Secondary	%	54	40	49	55	65	54	33	31	48	34	0	37
Digester SRT	D	38	20	27	40	27	28	14	16	21	19	22	22
Digester Feed VS	%	83	68	78	64	74	80	81	78	81	74	73	70
Digester Effluent VS	%	71	52	70	55	62	65	66	64	42	44	51	61
Digester Temperature	° C	37	36	36	36	36	34	37	52	37	36	36	36
Dewatering	Type	Low Solids Centrifuge	Low and High Solids Centrifuge	Low Solids Centrifuge	Lagoons	High Solids Centrifuge	Low Solids Centrifuge	Low Solids Centrifuge	High Solids Centrifuge	Plate and Frame	Low Solids Centrifuge	Low Solids Centrifuge	High Solids Centrifuge
Centrifuge Speed	RPM				N/A					N/A			
Conveyor	Type	Screw, Belt, Hopper, Truck	Screw, Belt, Truck	Belt, Silo		Hopper, Truck	Cake pump, Pad	Belt	Cake pump, Silo, Truck	Gravity into Truck	Belt	Belt	High Pressure Pump

[1] All plant information represents values averaged over the month prior to testing at the plant.

Table ES-3. Plant Field Analysis Data Collected by Project Team on Day of Sampling.

		Plant 1	Plant 2	Plant 3	Plant 4	Plant 5	Plant 6	Plant 7	Plant 8	Plant 9	Plant 10N	Plant 10S	Plant 11
Plant Flow	Mgd	36	200	68	30	20	60	17	350	13	350		200
Influent H_2S	ppmv	20	0.003		1.2	105	0.32	0.07	0.10	0.11	5.15		0.04
Influent ORP	Mv	-216			-75								147
Digester ORP	Mv	-142 -293	-251 -241	-210	-213 -201	-240	-268 -232	-307	-256	-39 +33	-188	-196	-175 -175
Digester off-gas H_2S	ppmv	2,100	1.3	11.4	250	15	0.31	2,000	200	90	25	25	1.5
Digester off-gas NH_3	ppmv	ND	<0.1	700	<0.1	0.25	6	ND	ND	2	1	2.5	<0.2
Digester off-gas CO_2[1]	%	38	33		38	38			36	35	33	35	33
Digester off-gas CH_4[1]	%	61	62		61	61			63	55	60	60	65
Dewatered Cake Temp	°C	34	31	37	22	27		36	49	33	32	32	39
Conveyed or Stored Cake Temp	°C	18	33	30	24	28		21	51	26	27	27	42
Stored Cake Temp	°C								36				26

[1] Digester off-gas analyzed by Performance Analytical Laboratories.

Table ES-4. Olfactometry Analysis Results for Post-Digestion Biosolids Samples, and Peak Sulfur and Nitrogen Values for the Same Samples.

Plant No	Sample Location	Odor Detection Threshold[1] (DT)	Odor Recognition Threshold[1] (RT)	Peak Total Sulfur mg S/m^3	Peak Total Nitrogen mg N/m^3
1	F2 Digested Biosolids	360	230	4.0	0
	G Fresh Low-Solids Centrifuge Biosolids Cake	17,000	11,000	2,0202	0
	I2 Low-Solids Centrifuge Cake after about 7-10 days storage	18,000	14,000	1,7743	0
2	F2 Digested Solids after Holding Tank	390	230	4.0	0
	G Low-Solids Centrifuge Biosolids Cake	6,100	4,300	352	0
	I High-Solids Centrifuge Biosolids Cake	21,000	13,000	787	0
3	F Digested Biosolids	460	270	4.8	0
	G Dewatered Biosolids	9,600	7,300	416	0
	I Stored Biosolids Cake	4,800	4,200	173	0
4	F2 Digested Sludge	230	120	5.0	0
	G Lagoon Top Biosolids Sample	3,700	1,600	60	0
	I Lagoon Top Biosolids Sample	3,500	2,000	27	0
5	F Digested Biosolids	270	140	2.7	0
	G High-Solids Centrifuge Biosolids Cake	6,100	3,500	494	1.1
	I High-Solids Centrifuge Cake after about 2 days storage	7,400	4,300	394	1.7
6	F1 Digested Biosolids (DS)	95	70	8.0	0
	F2 DS After Holding Tank	120	75	1.0	0
	G Fresh Biosolids Cake	5,100	3,100	139	1
	I Stored Biosolids Cake	2,900	1,700	131	1
7	F Digested Biosolids	1,300	830	19.3	0
	G Centrifuge Biosolids Cake	19,000	14,000	2,408[4]	4.34
	I Drying Bed Biosolids Cake	1,900	1,400	67	0.56
8	F2 DS Post-Screening	120	65	2.8	1.0
	G Centrifuge Biosolids Cake	9,100	5,000	621	2.23
	I1 Cake Post-Conveyance	2,500	1,400	578	3.33
	I2 Cake Post-Storage	8,900	6,100	304	2.85
9	F1 Digested Sludge	2,500	1,400	14.3	0
	F2 Digested Sludge Post-Storage	95	70	1.8	0

Table ES-4. Olfactometry Analysis Results for Post-Digestion Biosolids Samples, and Peak Sulfur and Nitrogen Values for the Same Samples. (cont.)

	G Plate & Frame Filter Press Cake	1,700	1,100	19	0.97
	I Cake Stored for Two Days	2,200	1,300	130	0.85
10	F2 Train #2 Digested Sludge	1,300	730	4.3	0
	G2 Train #1 Centrifuge Cake	12,000	8,100	874	0.91
	G3 Train #2 Centrifuge Cake	8,700	5,700	632	1.13
	I Stored Biosolids Cake	21,000	11,000	1,160	0.87
11	F2 Primary Digested Sludge	100	65	0.7	0
	G Centrifuged Biosolids	15,000	8,700	819	0.54
	I1 Cake Post-Conveyance	13,000	8,700	983	0.72
	I2 Cake Post-Storage	1,300	730	19	2.14

[1] These measurements were performed on gas samples obtained from headspace bottles on Day 6 of storage and collected in Tedlar® bags, without any headspace losses. Day 6 was chosen, since the days-to-peak values were not known at the time of sampling. The values represent a 1:50 dilution in samples.

[2] This value represents a 93% H_2S contribution in total sulfur concentration.

[3] This value represents a 89% H_2S contribution in total sulfur concentration.

[4] This value represents a 58% H_2S contribution in total sulfur concentration.

Table ES-5. Digester Performance Indicators.

		Plant 1	Plant 2	Plant 3	Plant 4	Plant 5	Plant 6	Plant 7	Plant 8	Plant 9	Plant 10N	Plant 10S	Plant 11
Labile Protein In	mg/g DS	48	58	24	26	24	41	47	41	43	15	15	40
Labile Protein Out	mg/g DS	14	20	20	13	16	19	15	44	28	15	18	18
NaOH Extracted Protein In	mg/g DS	231	207	459	259	286	310	242	261	185	90	80	222
NaOH Extracted Protein Out	mg/g DS	155	136	295	164	190	230	195	95	222	151	139	155
Digester VS Destruction	%	61	54	57	42	51	67	55	54	66	63	61	45
Dewatered Cake RBA FVSR		0.1471	0.1018 0.1344	0.1393	0.3139	0.1461	0.2092	0.0168	0.0621	0.0738	0.1844	0.1363	0.1873
Digester Total Fe	mg/L	1082	2200	485	1546	2173	884	233	619	745	2144	1438	1024

Following completion of the intensive field sampling and analysis period, the WERF project team compiled and examined the collected data to develop conclusions on the sources of odor generation at wastewater treatment plants. Based on these research findings, the project team developed conclusions and made recommendations for research on topics identified to be crucial for a better understanding of WWTP biosolids odor issues.

ES.7 Implications of Study Findings on Biosolids Odor Generation

With respect to all of the parameters analyzed in this study, biologically available protein was confirmed to be the main contributor to the odor potential of biosolids. As a result of biological activity in the anaerobic digesters, proteins are broken down into amino acids and are consumed as substrate by specialized microbial groups. These degradation reactions release the sulfur- and nitrogen-bearing volatile organic compounds as end products, and those end products are the main source of odors associated with biosolids. The protein measurements performed on digester influent and effluent, along with other familiar characterization parameters of the anaerobic digestion process are summarized in Table ES-5.

Because biosolids samples were collected and incubated (under ambient conditions) in small, air-tight sample bottles for at least six days, the olfactometry measurements in the bottle headspace were much higher than would be sensed by an individual onsite or from the olfactory analysis of an air sample gathered by flux chamber or another type of field headspace. The bottle-headspace method of sampling used in this project can simulate the odors from samples stored in bunkers, hoppers, or trucks and bound for land application or disposal. While the bottle-headspace sampling method may not represent actual field conditions, the consistency of sampling and analytical protocols that were used for all 11 WWTPs in the study results in a useful and statistically valid comparison of analytical results from the huge number of samples collected.

In the United States, the 503 Rule regulates production of biosolids and their beneficial use. The 503 Rule governs treatment and disposal or use of biosolids, based on compliance with pollutant limits and the degree of pathogen reduction (Class A or Class B). The rule also requires stabilization of treated biosolids, defined in the regulation as "vector attraction reduction" (VAR). The main goal of VAR is to achieve an acceptable degree of biological stability in biosolids by one of 12 VAR options recommended in Rule 503.

Although odors from biosolids are not specifically regulated by the 503 Rule, it is generally accepted that efficient anaerobic digestion and stabilization of biosolids in accordance with the 503 Rule should also lead to reduction of biosolids odors after digestion. However, a comparison in this study of classical digestion parameters, such as VS destruction or RBA, with odors from the dewatered biosolids cake did not reveal any conclusive results. While the 38% VS destruction requirement of VAR Option 1 was exceeded at all test WWTPs (VS reduction varied between 42% and 67%), the treated biosolids still produced relatively high odors and VSC emissions after centrifugal dewatering.

The project team found relatively long digester solids retention times (SRTs) at the WWTPs, varying from 14 to 40 days (75% of the WWTPs had digester SRT values greater than 20 days). Yet when examined in conjunction with VS destruction and dewatered cake odor emissions, the project team concluded that further study will be necessary to determine if any relationships exist between digester SRT, VS destruction, and biosolids odor emissions. Mechanistic studies should be undertaken to better understand the link between classical digestion parameters in terms of their contributions to odors in the final biosolids product from WWTPs.

ES.8 Conclusions and Recommendations

Summary conclusions of this study are as follows:

1. The anaerobic digestion process is not well understood with respect to its impact on odors from digested biosolids, especially with respect to odors in biosolids cake after centrifugal dewatering.
2. There is no conclusive evidence that more effective anaerobic digestion, leading to more complete digestion of complex organics, can result in lower odors in digested, dewatered biosolids cake.
3. There is further evidence that every step in a biosolids handling process has some impact on odor production from the digested biosolids cake. In other words, efficient digestion alone is not sufficient to maintain low odors when the upstream conditions or downstream processing equipment imposes conditions that increase the potential for odor generation.

Recognizing that this study finds a clear and conclusive relationship between bio-available protein and biosolids odor production, it follows that destroying more releasable proteins during digestion will result in lower odor production from biosolids cake. Therefore, processes capable of enhancing digestion to destroy higher amounts of VS and bio-available protein will result in lower odor emissions from biosolids cake. The project team concludes that a variety of pre-digestion processes for cell lysis, sonication, or pasteurization should be investigated for the enhancement of digestion efficiency. Some of these processes are the following:

- Pre-pasteurization
- Pre-digestion chemical treatment
- Thermal pre-processing
- Two-phase digestion (several possible types)
- Thermophilic anaerobic digestion (several possible types)
- Anaerobic digestion followed by aerobic digestion (mesophilic aeration)
- Mechanical cell-lysing techniques (sonication)

The project team recommends using controlled, mechanistic studies, which provide the capability to vary one input parameter to determine its cause and effect, to investigate further the impacts of mesophilic and thermophilic digestion on downstream biosolids odors. Also, appropriate processes need to be paired to investigate biosolids processing and handling combinations downstream of digestion, which should result in less odorous biosolids product.

Please note: Data files (Appendix A) for this research may be available through WERF and/or the third-party publishers listed on page ii and the back cover of this report. For more information, call the organization from which you purchased this report or contact WERF at 703-684-2470 or visit our website at www.werf.org.

CHAPTER 1.0

INTRODUCTION

1.1 Background

The authors of this report have been involved in a comprehensive odor control research project that seeks to identify and explore gaps in knowledge concerning odor issues that affect the wastewater treatment industry. Participating consultants and wastewater treatment agencies have provided more than $1.5 million of in-kind contributions of time and services for this project. The project is comprised of the following two major study phases:

1. A broad literature review, in which the WERF Project Team examined published and unpublished documents on odors associated with municipal wastewater collection systems and treatment facilities, including biosolids handling, and
2. A field and laboratory study to produce further research into the topic that was identified during the literature review as most needing investigation.

1.1.1 Phase I Literature Review on Wastewater Odors

In Phase I, the project team completed a comprehensive review of published literature and unpublished documents within the wastewater treatment industry, examining papers associated with nearly all WWTP processes including collection systems, pumping stations, wastewater treatment processes, and biosolids processing facilities. In addition, the project team reviewed the documented experiences of related industrial facilities and the agricultural industry, as well as "gray literature" describing the experiences of utilities in the wastewater treatment industry with respect to odors and odor control. The project team's report on Phase I of Project 00-HHE-5, *Identifying and Controlling Odor in the Municipal Wastewater Environment, Literature Search and Review*, was published by WERF in 2003.

Of the large number of documents reviewed in Phase I that related to biosolids odor issues, a few papers discussed odor emissions resulting from the anaerobic digestion of biosolids. According to these papers, many new anaerobic digestion systems are being put into operation and billions of dollars are likely to be spent on new anaerobic digestion facilities. There is a general belief in the wastewater treatment industry that anaerobic digestion reduces odor emissions by stabilizing the solids, but the project team did not find data in Phase I to support this conclusion.

During a specialty workshop at the 2000 Water Environment Federation Technical Exposition and Conference (WEFTEC) conference in Anaheim, Calif., the WERF Project Team presented a summary of the Phase I literature review and the recommendations for future research to more than 50 attendees. The attendees were asked to prioritize recommended research projects. They voted biosolids odor issues as the highest priority for future odor research. Specifically, the attendees wanted research on the influence of upstream treatment processes on biosolids odor quality. Storage, anaerobic digestion, bio-available protein, and polymer addition were parameters identified as potential influences on overall biosolids odor quality. The attendees also voted biosolids sampling and analytical measurements techniques as the fourth highest priority for future research. Thus, the majority of the attendees recognized the need for additional biosolids odor research.

1.1.2 Rationale for Phase II Field Study

Biosolids odor emissions can affect the ability of wastewater utilities to implement beneficial biosolids processing and reuse programs. Communities often become more sensitized and vocal about biosolids issues once they experience odors emanating from a nearby facility. Many national and local newspapers, citizens groups, and regulatory agencies have recently targeted odor impacts from biosolids, including potential human health effects. These groups have raised significant concerns about viable biosolids disposal methods and selection of disposal sites. Many national and local regulatory agencies are considering biosolids disposal bans in their communities because of misinformation, citizen fear, misperceived hazards, and potential odor impact concerns.

Odor is the number one factor that WERF member agencies cite for problems with biosolids disposal in their communities. The recently completed Phase I literature review and the workshop feedback concluded that the wastewater treatment industry has an incomplete understanding of the in-plant operations and treatment parameters that influence biosolids odor emissions. Without a better understanding of these factors, wastewater treatment plants (WWTPs) often have little control over the biosolids odor quality that is detected by local communities.

Many publicly owned treatment works (POTWs) have decided that anaerobic digestion will produce the least odorous biosolids product. However, relatively little odor data currently exist in wastewater treatment industry literature to support this belief, and there is insufficient data to show how anaerobic digestion system design and operating parameters can influence biosolids odor quality.

1.2 Purpose

In Phase II, the WERF project team collected data and drew correlations to demonstrate the influence of anaerobic solids digestion system design and operating parameters on the odor quality of the final biosolids product. Other biosolids stabilization approaches exist, but anaerobic digestion appears to be the most widely practiced operation in the United States on a weight percentage basis.

The project team determined how process conditions (storage, anaerobic digestion, and mechanical dewatering) affect odor emissions from biosolids and used established and new sampling and analytical methods to measure odor precursors in the liquid and gaseous phases of the biosolids, which were produced under a variety of set process conditions.

The project team drew correlations between the process conditions and the measured odor precursors to provide a better understanding of the conditions that produce more odorous biosolids. This research should help the wastewater treatment industry better manage odor quality and the resulting impacts on surrounding communities.

Eleven WWTPs participated in Phase II of the WERF Odor Study. The WWTPs are generically described and listed by code number in this report in order to protect their identities, the site-specific data, and the results.

1.3 Project Organization

The laboratories utilized in the project were: Virginia Polytechnic Institute and State University (VPI), Bucknell University, and St. Croix Sensory Laboratories. These laboratories provided offsite analysis of samples. The local laboratories of participating facilities provided additional analyses of wastewater and biosolids samples. The WERF project team also performed basic field measurements upon sampling and analyzed headspace samples for certain constituents on the day of sampling.

CHAPTER 2.0

SAMPLING AND TESTING PROCEDURES

To ensure a productive outcome, the project team developed a series of project guidance and quality control documents to guide samplers and reviewers through the overall sampling and analysis process.

The general testing approach involved collecting gas and liquid biosolids samples from the WWTPs participating in the WERF Odor Study. The project team collected design and operating parameters for the biosolids storage, anaerobic digestion, and dewatering facilities at the WWTPs. The samples were analyzed for certain measurable chemical and odor quality parameters. The project team then used statistical analyses to determine the degree to which the odor quality parameters correlate to the analytical data and/or design and operating parameters.

2.1 Sampling Procedures

Prior to testing at the WWTPs, the project team developed comprehensive testing and sampling procedures in a General Testing Protocol document. The General Testing Protocol encompasses system-operating parameters with monitoring/sampling points at logical locations throughout the process. Site-specific protocols were written for the individual WWTPs that took part in the study, all based on the General Testing Protocol. Figure 2-1 identifies the generic sampling locations within a WWTP biosolids process train that were analyzed as part of this study. The team identified corresponding sampling locations at each WWTP that took part in the program and added or subtracted locations specific to each's particular process schemes. In addition to these sample locations, the project team took a 24-hour composite sample of the influent to each WWTP.

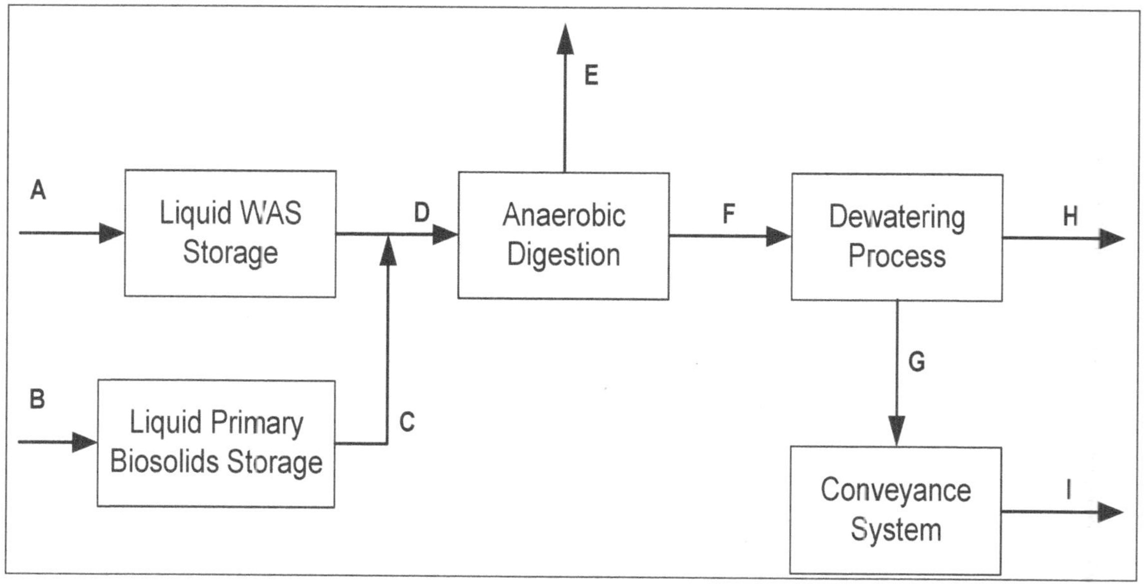

Figure 2-1. Generalized Test Facility Flow Schematic with Sample Locations.

The intent of the field study was to capture all biosolids treatment units upstream and downstream of anaerobic digestion and from the point where the biosolids leave the treatment facility to their ultimate disposal or reuse destination. The project team customized the general schematic diagram shown on Figure 2-1 for each WWTP to reflect its unique design and process conditions. Samples were collected before and after each of the biosolids treatment processes and analyzed for constituents in the liquid, solids, and gas headspace phases of each sample, as shown in Table 2-1. This large suite of samples provides information about changes in chemical and odor characteristics of biosolids as they proceed through the biosolids treatment train.

Table 2-1. Sampling and Analyses Matrix as a Function of Sample Location.

Analysis	Sample Location									
	Influent	**A**	**B**	**C**	**D**	**E**	**F**	**G**	**H**	**I**
Sample Type	Liquid	Liquid	Liquid	Liquid	Liquid	Gas	Liquid	Cake	Liquid	Cake
Standard Tests (performed by facilities)[1]										
Solids (VS & TS)	X	X	X	X	X		X	X	X	X
ORP/pH/ temp	X	X	X	X	X		X	X	X	X
Alk/NH_3/ TKN	X	X	X	X	X		X	X	X	X
VFA/ Coliforms							X			X
VS Destr/ Methane					X	X	X			
Field Odor Tests (onsite by project team)										
Colorimetry H_2S						X				
Jerome H_2S	X	X	X	X	X		X	X	X	X
Colorimetry NH_3	X	X	X	X	X		X	X	X	X
Headspace Analysis (VPI)[3,4]										
Reduced sulfur	X	X	X	X	X		X	X	X	X
Amines	X	X	X	X	X		X	X	X	X

Table 2-1. Sampling and Analyses Matrix as a Function of Sample Location.

Analysis	Sample Location									
	Influent	A	B	C	D	E	F	G	H	I
Ketones	X	X	X	X	X		X	X	X	X
Olfactometry[5]							X	X		X
Organic Compounds Analysis (Bucknell)[2]										
Proteins			S,T	S,T	S,L,T		S,L,T	L,T	S,T	L,T
Enzyme Assays					L		L	L		L
Amino Acids					T		S,L,T	L		L
VFA				S	S		S			L
Cations & Anions (Bucknell)[2]										
Fe/Al	T				T		T		T	
Ca/Mg/Na/K	S	S			S,T		S,L,T	L	S	L
Sulfate			S	S	S					
Sulfide							S,L	L		L
Residual Biological Activity (RBA) (VPI)[2]										
Additional VS Destruction							X	X		X
Methane							X	X		X
Ammonia (NH_3): the extraction yields soluble + bound NH_4							T	T		T

Notes:

1. X = Sampling and analysis point at that location
2. Employ analysis of fractions: S = soluble, L = labile, T = total NaOH extracted
3. Repeated analysis of samples from Sample Locations F, G, and I, every day for 1 week of lab incubation
4. Measurement of analytes that require longer gas chromatography times were conducted 48 hours after sampling and onwards for samples obtained from Sample Locations F, G, and I only.
5. Analysis by St. Croix Sensory Labs on Day 7 after sampling at VPI at Day 6 of lab incubation

The project team loaded 141 grams (g) of biosolids into empty 500 milliliter (mL) polyethylene terephthalate (PET) bottles (Aquafina, Pepsi Cola Company). The bottles were sealed and incubated for up to 49 days at 22°C prior to odor analysis. Table 2-2 shows the sample preparation schedule for a typical facility test site. Approximately 150 sample bottles were prepared for each WWTP, in addition to the samples collected and prepared for analysis of standard methods by each WWTP's local laboratory.

Table 2-2. Breakdown of Sample Bottle Preparation for Each Test WWTP.

Sample Location	Type of Analysis						
	Field Odor H_2S (Jerome)	Field Odor NH_3 (ASTM D4490)	Headspace (VPI)	Olfactometry (ASTM E679) (VPI)	RBA (VPI)	Organic Compounds, Cations, and Anions (Bucknell)	Container Volume (Bucknell Samples)
Influent	2	2	4			2	500 mL
A	2	2	4			2	100 mL
B	2	2	4			2	250 mL
C	2	2	4			2	250 mL
D	2	2	4			2	1L
E	See Note 2						
F	2	2	12	2	6	2	2 L
G	2	2	12	2	6	2	1 L
H	2	2	4			2	1 L
I	2	2	12	2	6	2	1 L
Duplicates	NA	NA	4	1	2	2	Total (below)
Total	**18**	**18**	**64**	**7**	**20**	**20**	**147**

Notes:

1. Samples collected for Field Odor Tests involving the Jerome meter and NH_3 did not require sample bottles to be shipped to laboratories (numbers shown indicate the number of tests run in the field).
2. The digester gas H_2S sample at Sample Location E was composited in one Tedlar® bag and measured in the field using a colorimetric tube rather than the Jerome meter.

2.2 Analytical Rationale

Based on prior research, the putrefaction or odor production potential of anaerobically digested biosolids is related to the presence of residual biological activity (RBA) or substrate. Figure 2-2 shows the process locations within a typical WWTP on which this study focused to explore this hypothesis and some of the mechanisms related to these unit processes that are thought to result in the production of odors from biosolids.

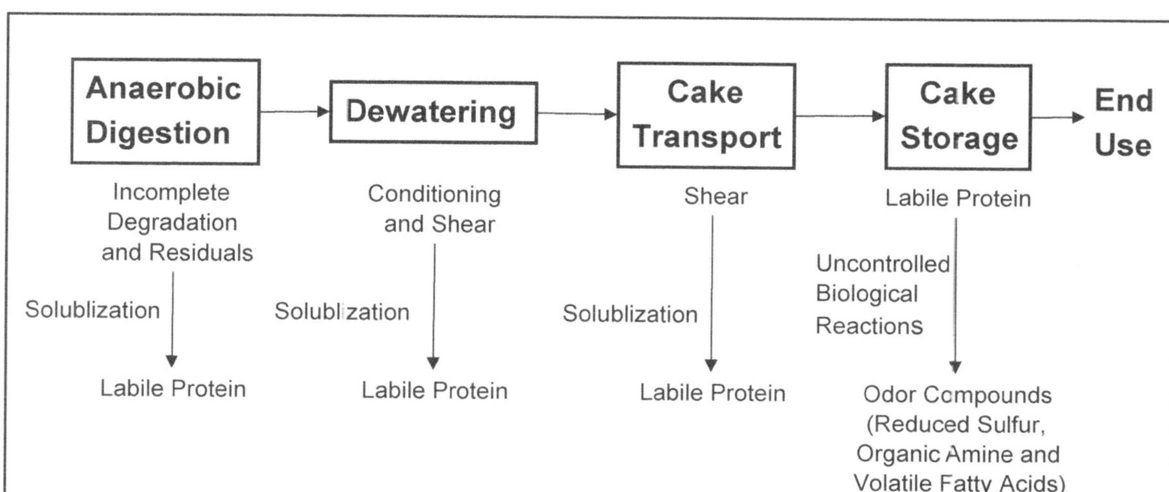

Figure 2-2. Theoretical Pathways for Odor Production from Biosolids.

The presence of residual substrate may be a result of digester performance, which may be affected by process parameters such as detention time, temperature, mixing, volatile solids (VS) destruction, alkalinity, ammonia (NH_3), and gas production. These process parameters have been obtained from the digesters that were sampled in this study. The presence of degradable substrate, such as labile protein, was also measured using colorimetric techniques. The analysis of RBA, such as additional VS destruction, was performed in the laboratory by measuring the end products of biological reactions, such as methane or NH_3.

2.3 Analytical Procedures

The sample analyses performed by each of the participating laboratories are described in more detail below.

2.3.1 Standard Tests of Water and Wastewater Analyses

The WWTPs analyzed liquid samples for VS, total solids (TS), temperature, pH, ORP, sulfides, alkalinity, total Kjeldahl nitrogen (TKN), and fecal coliform bacteria. All tests in this subsection were completed under Standard Methods (American Public Health Agency, 1999) procedures, as referenced below. Examples of methods used by the WWTPs include the following:

- VS: Method 2540D
- TS: Method 2540E
- ORP: Method 2580B
- Temperature: Method 2550B
- pH: Method 4500-H+B
- Alkalinity: Method 2320B
- TKN (soluble): Method 4500-N_{org}B
- Sulfides: Method 4500-S^{2-}B
- Fecal Coliforms: Method 9222B

The team used the analytical results provided by these Standard Methods, along with the process and design data from each WWTP tested, to provide the fundamental basis by which the WWTPs were compared.

2.3.2 Field Analyses

In the field, a team member measured gas-phase hydrogen sulfide (H_2S) in the headspace of the digester gas at Sample Location E and all post-digestion samples. All measurements of H_2S, with the exception of digester gas H_2S, were completed using a Jerome 631-X Gold Film H_2S Analyzer. While the Jerome 631-X H_2S Analyzer is subject to interference by other reduced sulfides and, therefore, is not an exact measurement of H_2S, its measurements are sufficiently accurate and useful for comparison on a relative basis. A project team member also measured gas-phase H_2S concentrations within the digester gas at Sample Location E using a colorimetric tube standard, where the gas is collected in the field inside a Tedlar® bag.

A project team member analyzed gas-phase NH_3 concentrations in the field using colorimetric tubes per American Society for Testing and Materials (ASTM) D-4490. Tubes were available for concentration ranges of 0.25 to 500 parts per million by volume (ppmv).

2.3.3 Headspace Analysis of Odorous Chemical Compounds and of Odor (Olfactometry)

The project team used a combination of headspace analysis of odorous chemical compounds and analysis of odor (olfactometry) to establish a relationships between major odorous chemical compounds and the human perception of odor. Headspace gas sample of the same sample was used to measure and to compare the concentration of odorous chemical compounds, the concentration of odor units, and the quality/nuisance of odor.

2.3.3.1 Procedure and Rationale of Chemical Headspace Analysis of Odorous Chemical Compounds

The project team used headspace gas chromatograph/mass spectrophotometer (GC/MS) analyses on samples in closed bottles to estimate differences in odor emissions of different biosolids over 49 days, wherein it was possible to track the distinct dynamics of production and transformation of sulfur odor compounds.

The project team performed headspace analyses on the gas-phase of the samples taken from sealed 500-mL bottles (PET-polymer, gas-tight, inert, used as an incubator at 22°C) that contained 141 g of biosolids, by inserting the airtight gas sampling syringe needle directly through the bottle cap. Each sample was used for only one analysis, then discarded. For WWTPs No. 1 through No. 5, the extracted headspace sample was analyzed using a Hewlett Packard (HP) GC/MS with a 30 m x 0.32 mm internal diameter (ID) x 0.25 mm HP 5 column. The carrier gas flow was 2 mL/min, with a sample size of 1 mL, and the temperature in the cryo-concentration unit of –130°C for the first loop of the GC column. For WWTPs No. 6 through No. 11, 100 mL of headspace gas from the sample bottles was analyzed by GC/MS running at the same 2 mL/min carrier gas flow rate. Additionally, the boiling point span of the analyzed headspace components ranged from – 88°C to + 280°C in each GC/MS run, and the column used was 30 m x 0.32 mm ID x 1 mm PDMS 5% Phenyl (Agilent). Data acquisition was achieved by running the GC/MS in "scan" mode.

For quantification, standards were injected daily before measurements. The detection limit was on the order of 100 parts per billion by volume (ppbv), and the precision of repeated analysis was on the order of ±5%. Analytes measured included hydrogen sulfide, methane thiol, dimethyl sulfide, dimethyl disulfide, dimethyl trisulfide, carbonyl sulfide, carbon disulfide, carbon tetrachloride, isopropyl mercaptan, tert-butyl mercaptan, n-propyl mercaptan, ethyl methyl sulfide, trimethyl amine, acetone, indole, and skatole.

The headspace sampling method used in Phase II has several advantages over flux chambers. The storage conditions in headspace bottles are anaerobic and mirror the anaerobic bulk core of large full-scale cake piles. The odor concentration in the headspace of the bottles is a function of the odor concentration contained in the cake.

The headspace storage bottles, which are closed bioactive systems, allow the measurement of odor consumption by the cakes in the bottle and the odor production/ consumption cycles. As a major advantage over flux devices, the headspace method is a simple means of deodorizing sulfur by two weeks of storage before transport (odor production/

consumption cycles of biosolids). The headspace method used during this project, combined with cryotrap-GC/MS, multiplied the productivity, speed of odor data production, and the quality of odor data, compared to flux chamber/SPME methods.

The headspace method is useful not only for chemical odor analysis (gas chromatography, test tubes, sensors can all be adapted to headspace bottles), but also for comparing chemical analysis with olfactometric odor analysis using human test panels (WWTP sending headspace bottles instead or in addition to Tedlar® bags).

2.3.3.2 Procedure and Rationale of Olfactometry Headspace Analyses

The olfactometry samples were produced by 1 to 50 dilution of the same headspace gas used for chemical headspace analysis on Day 6 of sample incubation in bottles. Olfactometry was conducted in accordance with ASTM Standard Practices (ASTM, 1991; ASTM, 1999), using a trained odor panel. The olfactometry produced the following results:

- Odor Thresholds, defined as Detection Threshold (DT) and Recognition Threshold (RT), were determined using the "triangular forced-choice dynamic dilution" presentation method, as prescribed by ASTM E679-91. Results are expressed in dilutions-to-threshold (D/T) units or Odor Units (D/T), as described below.
- Odor Descriptors: The odor panelists described specific odor characteristics of the samples, such as sour, earthy, ammonia-like, and fishy.

The standard olfactory analysis for each site included DT, RT, and odor descriptors for each sample. The purpose of the project was to compare the chemical odor measurement of significant odor compounds (sulfur and nitrogen) using a GC/MS with olfactory odor measurement by using human test panels. The samples for olfactory measurement were produced under exactly the same conditions as the GC/MS samples, but only for Day 6 of headspace incubation.

On Day 6 of incubation headspace gas (100 mL headspace, diluted to 5 L with nitrogen in a Tedlar® bag) from biosolids bottles was taken for olfactometry and sent overnight to St. Croix Sensory Labs, Inc. for analysis on Day 7 (Day 0 starts when biosolids samples are loaded into the bottles and sealed on plant site). St. Croix Sensory Labs conducted olfactometry analysis in accordance with ASTM International and European Committee for Standardization (CEN) standards. The analyses included the determination of odor detection and recognition thresholds and odor character profiling.

Odor thresholds were determined using triangular forced-choice olfactometry, following ASTM International E679-91, "Standard Practice for Determination of Odor and Taste Thresholds by a Forced-Choice Ascending Concentration Series Method of Limits" and the European Union testing method EN13725, "Air Quality – Determination of Odor Concentration by Dynamic Olfactometry." This testing was performed with the AC'SCENT® International Olfactometer, which is a dynamic dilution olfactometer operating at a presentation flow rate of 20 L per minute.

The dilution ratio result of olfactometry, known as the DT, is an estimate of the number of dilutions needed to make the actual odor emission just detectable. RT is the dilution ratio at which the assessor firsts detects the odor's character ("smells like").

Assessors also observe the odorous air samples at full strength from the sample bag and report odor descriptors for odor character profiling of the samples. Assessors utilize a character profile reference vocabulary for selecting the odor characters from eight main categories: floral, fruity, vegetable, earthy, offensive, fishy, chemical, and medicinal. A histogram of character descriptors reported by assessors was provided for each odor sample evaluated.

All assessors used for this project were selected and trained following ASTM International STP 758, "Guidelines for Selection and Training of Sensory Panel Members" and selection and training criteria specified in EN13725.

2.3.4 Organic Compounds Analyses–Extractions

Liquid-phase analyses of organic compounds were performed on the samples collected per Table 2-1, including the measurement of proteins, amino acids, enzyme activity, and volatile fatty acid (VFA). For each of these organic compounds, the analysis was further divided into fractions, which include soluble, bound or labile, and total organic constituents.

Samples were filtered to analyze for soluble organic fractions within the samples. To obtain a filtered sample for measuring the solution properties, an appropriate volume of sample was filtered through a 1.5 µm filter. The sample was centrifuged in a laboratory centrifuge at 10,000 gravity-force units (Gs) for 15 minutes.

Biosolids extraction was performed to analyze for the bound organic compounds within the samples. The solids pellet from the laboratory centrifuge (or in the case of biosolids cake samples, 10 g of cake) was re-suspended in pH 8 phosphate buffer saline (PBS solution). Distilled water was used in place of PBS for extraction of VFAs from the cake only. The bound or labile fraction from this fraction was extracted using 10 minutes of shear at G=1000/s. The suspension was centrifuged at 3,000 Gs for 15 minutes at 5°C, and the supernate was filtered using a glass microfiber filter (1.5 µm). This fraction was labeled as the "labile fraction."

Total sample analysis for organic compounds was conducted as indicated in Table 2-1. Samples were washed and suspended in 1 Normal (N) sodium hydroxide (NaOH) for total protein analysis. This mixture was stirred by a magnetic stirrer at approximately 500 revolutions per minute (rpm) for two hours. To obtain a sample for analysis, an appropriate volume of sample was filtered through a 1.5-µm filter after centrifuging the liquid sample in a laboratory centrifuge at 3,000 Gs for 15 minutes. Although this digestion method does not extract all protein from the samples, for simplicity the protein extracted is referred to as "NaOH extract" protein.

2.3.5 Protein, Enzymes, and Acid Analyses

Protein was measured in three fractions: soluble proteins, labile proteins, and NaOH extracted protein. These fractions were obtained as described above. Protein concentration was quantified using the Hartree (1972) modification of the Lowry et al. (1951) method. Bovine serum albumin was used as the protein standard.

Enzyme assays were conducted for the labile fraction only of samples taken from Sample Locations D, F, G, and I (Figure 2-1). Leucine-aminopeptidase activity was measured using a modification of the method described by Teuber and Brodisch (1977). L-methionine-γ-lyase (METase) activity was measured according to the method of Yoshimura et al. (2000) with a slight modification using 1-methionine p-nitroanilide as the enzyme substrate and continuous detection of absorbence at 405 nanometers (nm) using a Thermo-Spectronic, Biomate UV/Visible Spectrophotometer. All enzyme assays were performed at 30°C.

Individual amino acids were measured using High Performance Liquid Chromatography (HPLC) subsequent to derivation procedures. Total amino acids were measured from Sample Locations D and F, soluble amino acid fraction at Sample Location F, and the bound fraction at Sample Locations F, G, and I (see Figure 2-1). The following steps were performed for amino acids:

1. The sample was oxidized with performic acid to convert cysteine and methionine to methionine sulfone and cysteic acid, which allows for better recovery of sulfur-containing amino acids.
2. The samples were hydrolized in 6 N hydrochloric acid (HCl) with 1% phenol and dissolved in sodium citrate buffer for 24 hours at 110°C.
3. Amino acid analysis was performed after hydrolysis of the protein samples with hydrochloric acid. The amino acids were derived and separated with HPLC and UV detection.

VFAs were measured in the liquid phase of the samples for the sampling points shown in Table 2-1. VFAs were measured on the soluble fraction of samples from Sample Locations C, D, and F and the bound fraction from samples taken at Sample Location I (Figure 2-1).

The C-2 through C-7 VFAs were measured using a HP 6890 GC with a flame ionization detector (FID). The column was a 0.53 mm ID, 30 m, Nukol capillary column (Supelco). The injector and FID temperatures were both 200°C. The column temperature was 110°C for the first five minutes, followed by a temperature ramp of 10°C/min to a final temperature of 150°C. The carrier gas was nitrogen at a flow rate of 20 mL/min.

2.3.6 Cations and Anions Analyses

Cations and anions were measured in the liquid-phase of the samples for the sampling points indicated in Table 2-1. Similar to the organic compounds analyses, the cations and anions samples were divided into fractions, including soluble, bound or labile, and total amounts of the various constituents.

2.3.6.1 Cation Analysis

The total metals content, including Fe, Al, Ca, Mg, Na, and K, was determined for samples from Sample Locations Influent, D, F, and H. The analysis required nitric acid-heat digestion to dissolve the particulate bound metals, followed by analysis using inductively coupled plasma (ICP). Method 3030 E in Standard Methods (1999) was used for the nitric acid-heat digestion procedure. The digest was measured by ICP according to Method 3500. The cations Ca, Mg, Na, and K were measured on the soluble and/or labile fraction from the Influent and from Sample Locations A, D, F, G, H, and I, as shown in Figure 2-1. Filtered samples were analyzed for cation concentrations using a Dionex Ion Chromatograph with a CS12 column and conductivity detection, with self-generating suppression of the eluent. The eluent was 20 millimolar (mM) methane sulfonic acid, introduced at a flow rate of 1mL/min.

2.3.6.2 *Anion Analysis*

The anion analysis included determination of sulfate and sulfide quantities at select locations in the soluble and/or bound forms (Figure 2-1). Soluble-, labile-, and NaOH-extracted fractions were determined the same way as for the organic analysis. Sulfate was measured on the soluble fraction of Sample Locations B, C, and D. Sulfate was analyzed by ion-chromatography using electronic self-regenerating suppression of eluent conductivity and conductimetric detection. The eluent is a mixture of 1.8 mM sodium carbonate and 1.7 mM sodium bicarbonate with a flow rate of 1.5 mL/min. The column is a Dionex Ion-Pac AS4A-SC with an AG4A guard column.

Sulfide analysis was performed on the labile fraction from the cake samples at Sample Locations G and I, in addition to the soluble fraction of the sample taken from digested solids at Sample Location F. Sulfide was measured using the Iodometric Method in Standard Methods (1995).

2.3.7 Residual Biological Activity Analyses

Additional VS reduction (AVSR) and methane and NH_3 production for liquid or cake samples were determined from sampling points indicated in Table 2-1. Liquid or cake anaerobically digested biosolids were batch-digested in a bench-scale unit for 40 days at 35°C, as specified in the 503 Rule by Option No. 2 for VAR, to evaluate RBA. The analysis for VS destruction, methane, and NH_3 occurred after one day and 40 days of arrival at the laboratory (with days counted from arrival at the laboratory, the day after completion of sampling).

2.3.7.1 *Estimation of RBA by Additional Volatile Solids Reduction (AVSR)*

VS destruction was evaluated using the Van Kleeck formula, which was applied to non-flow batch-fermentation in bottles (Appendix C of U.S. Environmental Protection Agency, 1999) using samples taken from Sample Locations F, G, and I. The weight of a 35-g sample stored in 500-mL sample bottles was determined for wet, dry, and inert content after one day and 40 days of storage, and evaluated for AVSR after 40 days of storage.

2.3.7.2 *Headspace Methane and Carbon Dioxide RBA*

Methane and carbon dioxide (CO_2) concentrations were measured using a GC thermal conductivity detector (TCD) by injecting a certified gas standard of methane and CO_2. The following protocol was used:

1. Sampling was done with 35-g wet biosolids (liquid or cake) samples in 500-mL sample bottles.
2. Pressure was controlled by controlled leak (small hole in bottle screw cap sealed by plastic tape).
3. Two bottles per day (sacrificed) were analyzed after one day and 40 days of storage, according to standard procedure.
4. Quantity (volume) of gases was measured by multiplying the mass of lost biogas (by measuring the mass loss of the bottles at Day 40) with a factor to convert gas-mass into gas-volume.

2.3.7.3 Ammonium Ion RBA

Analysis for ammonium ions was performed on the bound fraction of samples taken from Sample Locations F, G, and I. Samples from Sample Location F were centrifuged and passed through a 0.5-micron filter before being measured in the liquid phase. For Sample Locations G and I (cake samples), the bound fraction was analyzed after water addition and extraction of a centrifuged and filtered aquatic extract. An ion-chromatograph DX-300 from DIONEX and a DIONEX application was used, and the column was a CG16 and CS16 cation exchange column set at 20°C. The effluent was methanesulfuonic acid. Quantification was made by a standard of ammonium chloride.

CHAPTER 3.0

WWTP DESCRIPTIONS AND TEST RESULTS

The project team scheduled the sampling events for the field testing program during the summer when biosolids odor problems are typically the worst. A total of 11 WWTPs were sampled between May 22 and August 23, 2002. Most of the WWTPs are located in urban areas near the east and west coasts of the United States. Ten of the test facilities are located in the United States and one test facility is located in Canada.

The capacities of the participatingWWTPs ranged from 13 mgd to over 350 mgd, with five test facilities having capacities of 30 mgd or less and six test facilities having capacities over 100 mgd. The WWTPs received predominantly domestic wastewater flows, with varied mixes of industrial contributions, which is typical of large urban areas. All of the WWTPs employed anaerobic digestion, with 10 of the digestion systems operating in the mesophilic temperature range and one operating in the thermophilic temperature range. The WWTPs employed various technologies for biosolids thickening, dewatering, conveyance of dewatered cake, and end use or disposal.

Members of the project team conducted a one-day site visit at each WWTP to collect information for the development of site-specific protocols. A primary component of the site-specific protocols was determining the sample locations that best matched the general locations indicated in the General Testing Protocol. The site-specific protocol schematic diagrams use the same format as the overall General Testing Protocol schematic (Figure 2-1), but include the location and descriptions of specific process units at the facility and take into account additional process schemes and unique aspects of solids treatment and conveyance.

The site-specific protocols also include a customized sampling plan applicable to the facility. The project team strictly adhered to the customized sampling plans during testing. Since most of the laboratory analyses were performed in the week after sampling, the WWTP sampling events were set at least eight days apart. Sampling and analytical work continued through September 2002.

During a trial run sampling event, the project team determined that samples taken at locations upstream of the digesters were somewhat variable in measured physical parameters (pH, temperature, ORP), whereas samples taken from the digesters and locations downstream were consistent throughout the sampling day. Therefore, the project team prepared composite samples from four grab samples taken from locations upstream of the digesters over the course of a morning (Figure 2-1, Sample Locations A–D). The project team used grab samples from the digesters and further downstream locations (Figure 2-1, Sample Locations E–I).

During the site-specific protocol development, the project team also requested historical design and operational data from the test facilities to be incorporated into the data analysis portion of the project. Prior to all testing, the project team sent a Request for Information (RFI) letter to each facility, indicating the process and operational parameters of interest for the study. Following testing of all facilities, the project team compiled specific data in spreadsheet files for each facility. Appendix A of this report, which contains the detailed RFI spreadsheets, is available from WERF.

Each RFI spreadsheet contains general, process, and operational data for the facility averaged out for the week and month prior to the testing date. The following are examples of data acquired from the facilities and inserted into each RFI document:

- Influent flow, biological oxygen demand (BOD), TKN, and total suspended solids (TSS)
- Primary sludge flow, TS concentration, and VS content
- Activated sludge system mean cell residence time, aeration basin volume, and return activated sludge (RAS) rate
- Waste activated sludge (WAS) flow, TS concentration, and VS content
- Thickened sludge flowrates and concentration
- Digested sludge flow, TS concentration, and VS content
- Dewatered biosolids concentration and production (dry tons per day)
- Onsite storage detention time for dewatered cake

After insertion of the operational and process data into the RFI spreadsheet, a number of process parameters were calculated and later used to develop correlations with the analytical results. Following are examples of process parameters calculated:

- Primary and WAS mass quantities in wet and dry tons per day
- Solids detention times in the primary clarifiers, activated sludge processes, primary and secondary storage tanks, thickeners, and anaerobic digesters
- Total flow to anaerobic digesters
- Wet and dry tons of solids fed to anaerobic digesters per day (TS and VS)
- Wet and dry tons of solids removed from the anaerobic digesters per day (TS and VS)
- Digester VS destruction
- Wet and dry solids density for the dewatering facilities and cake production in wet tons per day

Table 3-1 contains a summary of information gathered from the participating agencies about the WWTPs taking part in this project. The following subsections present descriptions of each WWTP that participated in the study, including the facility size and wastewater influent characteristics; liquid and solids treatment processes; and other pertinent information.

Table 3-1. Summary of the Information Gathered from the RFI Forms Provided by the Agencies

	Plant Flow	CAS or HPO	Domestic Wastewater Contribution	WAS Thickening	Digestion Type	Digestion SRT	Dewatering Type	Dewatered Biosolids TS Content	Chemical Addition Locations		Biosolids Conveyor Type
									Iron	Polymer	
	mgd		%			D		%			
WWTP No. 1	36	HPO	100	DAFT	Mesophilic	38	Low Solids Centrifuges	18		DAFT, Dewatering	Screw, Belt, Hopper, Truck
WWTP No. 2	200	HPO	90	DAFT	Mesophilic	20	Low & High Solids Centrifuges	28 (Low) 35 (High)			Screw, Belt, Truck
WWTP No. 3	68	HPO	90	DAFT	Mesophilic	27	Low Solids Centrifuges	23	Dewatering	DAFT, Dewatering	Belt, Silo
WWTP No. 4	30	CAS	60	Belt	Mesophilic	40	Lagoons	5	Grit Channel	Belt, Secondary Clarifiers	
WWTP No. 5	20	CAS	90	Belt	Mesophilic	27	High Solids Centrifuges	24	Aeration Effluent	Belt, Dewatering	Hopper, Truck
WWTP No. 6	60		90	Gravity Thickener	Mesophilic	28	Low Solids Centrifuges	21		Belt, Dewatering	Cake pump, Pad
WWTP No. 7	17	CAS	95	Gravity Thickener	Mesophilic	14	Low Solids Centrifuges and Drying Beds	27		Dewatering	Belt
WWTP No. 8	350	HPO	85	Centrifuge	Thermophilic	16	High Solids Centrifuges	33	Primary Effluent	Headworks, Dewatering	Cake pump, Silo, Truck
WWTP No. 9	13	CAS	91	DAFT	Mesophilic	21	Plate and Frame	26	Communitor	Grit Removal, Secondaries Clarifiers, DAFT Plate and Frame	Gravity into Truck
WWTP No. 10	350	HPO	80	DAFT	Mesophilic	19	Low Solids Centrifuges	27 (North) 25 (South)	Primary Sludge	Belt	Belt
WWTP No. 11	200	CAS	93	DAFT	Mesophilic	22	High Solids Centrifuges	25	Pre-Primary	DAFT	High Pressure Pump

The subsections also present the site-specific protocol sampling schematic diagram for each facility. A general summary of field testing results is provided, including temperature, pH, and ORP of the samples, as well as the field headspace testing results for H_2S and NH_3. The tables indicate the testing results for odor at each facility. Finally, the following subsections present general observations and unique features of each testing event.

3.1 WWTP No. 1

3.1.1 WWTP Description

WWTP No. 1 is a medium-sized facility that treats wastewater generated in a mostly suburban, coastal area with a fairly mild climate. The facility treats domestic wastewater, with a significant commercial contribution from restaurants, hotels, and offices. There is no significant industrial discharge to this facility.

The nominal capacity of the WWTP is approximately 36 mgd. On the day of testing, May 22, 2002, the influent flow was 29 mgd. This facility is a high-purity oxygen, activated sludge plant with primary and secondary clarification. Sodium hypochlorite is used for effluent disinfection. Figure 3-1 depicts the solids flow schematic for WWTP No. 1, along with each sample location.

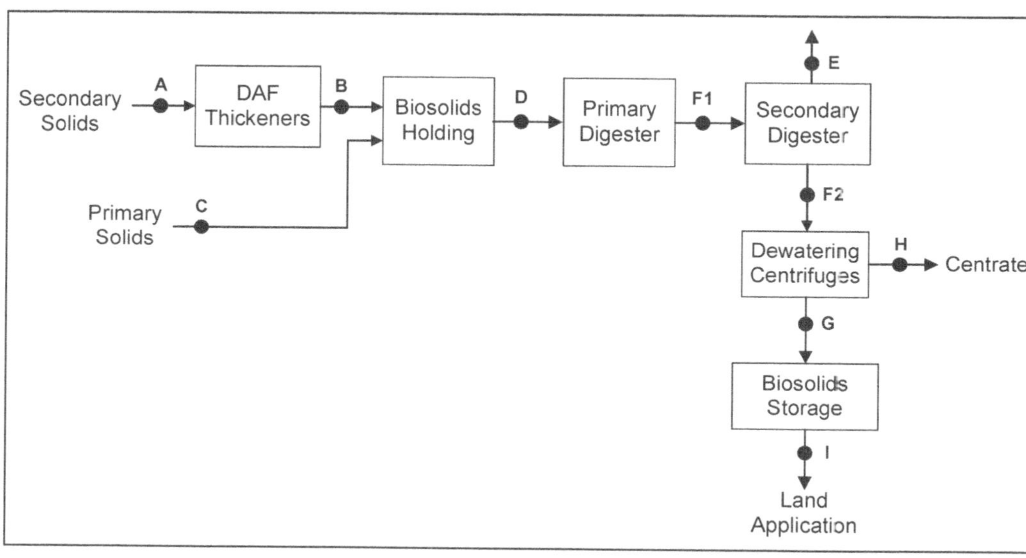

Figure 3-1. Schematic of WWTP No. 1.

Prior to anaerobic digestion, primary solids (PS) are removed from the rectangular primary clarifiers and pumped to one of four combined solids storage tanks ("Biosolids Holding" in Figure 3-1). WAS is thickened to about 4.5% total solids by four dissolved air flotation (DAF) units (three units were in operation on the day of sampling) and stored in small holding tanks (one per DAF unit). The WAS is then pumped into the four combined solids storage tanks (one per digester train) and mixed with the PS.

From the pre-digestion storage tanks, the combined primary and thickened WAS (TWAS) is pumped into four primary anaerobic digesters operating in the mesophilic mode and sized for 18-day detention time at design capacity. From there, the solids flow into one of two secondary gas-holding digesters with an eight-day detention time at design capacity. Digested biosolids are pumped directly from secondary digesters to three identical solid-bowl centrifuges of low-solids type for dewatering to a biosolids cake with approximately 18% dry solids content.

The dewatered biosolids are conveyed outside by a belt conveyor system to the top of a biosolids cake storage hopper with a capacity of approximately 150 wet tons. Biosolids from this hopper are discharged directly onto a truck bed and hauled a short distance to a covered concrete pad for storage prior to land application. The concrete pad has sufficient storage capacity for 60 to 90 days of cake biosolids production, depending on stacking height.

3.1.2 General Summary of Results

Table 3-2 presents the results of the field tests on samples collected at Sample Locations A through I at WWTP No. 1.

Table 3-2. Field Testing Results from WWTP No. 1.

Sample Location	Composite or Grab	Temperature (°C)	pH	ORP (mV)
INF Influent	Composite	22	7.0	-216
A WAS	Composite[1]	22	6.4	-115
B TWAS	Composite[1]	22	6.3	-177
C1 Liquid PS	Composite[1]	22	5.5	-192
D Combined PS and WAS	Composite[1]	22	5.5	-196
F1 Primary Digester Sludge	Grab	37	7.1	-142
F2 Secondary Digester Sludge	Grab	35	7.1	-293
G Dewatered Biosolids Cake	Grab	34	8.5	-55
H Centrate	Grab	34	7.5	-183
I Stored Biosolids Cake	Grab	18	8.1	-50

[1] Three composite samples, separated by 1 hour (results are average of three composites)

Data from the headspace field testing at WWTP No. 1 are presented in Table 3-3. The values shown represent averages of all samples taken, including duplicate samples. Correlations between these results and plant parameters are presented in Chapter 4.0 of this report. Hydrogen sulfide headspace concentrations were measured with a Jerome analyzer or sensory tubes. Because hydrogen sulfide is cross-sensitive with other reduced sulfur compounds, results should be considered a relative indicator and not absolute.

Table 3-3. Field Headspace Testing Results from WWTP No. 1.

Sample Location	H_2S Concentration (ppmv)	NH_3 Concentration (ppmv)
INF Influent	20	0.5
A WAS	0.17	3.5
B TWAS	1.2	0.7
C1 Liquid PS	350	0.2
D Combined PS and WAS	70	0.7
E Digester Gas	2,100	0
F1 Primary Digester Sludge	10	65
F2 Secondary Digester Sludge	30	60
G Dewatered Biosolids Cake	1.1	40
H Centrate	0.3	40
I Stored Biosolids Cake	1.0	35

Table 3-4 presents the results of the odor analysis on samples collected for WWTP No. 1. On May 29, 2002, one week after sampling, St. Croix Sensory measured odor intensity in the headspace gas from all sampling points downstream of digestion, with the exception of Sample Location H (dewatering centrate). The headspace samples were diluted 1:50 with nitrogen gas before shipping to St. Croix Sensory. Therefore, actual detection and recognition thresholds from the undiluted headspace samples were 50 times what is presented in Table 3-3. The sample results from diluted headspace samples are presented here so there is no confusion when they are compared with the lab results from St. Croix Sensory. Detection and recognition thresholds were measured as part of the olfactometry analysis.

Table 3-4. Odor Evaluation Results from WWTP No. 1.

Sample Location	Detection Threshold (D/T)	Recognition Threshold (R/T)
F2 Digested Biosolids	360	230
G Fresh Low-Solids Centrifuge Biosolids Cake	17,000	11,000
I 2 Low-Solids Centrifuge Cake after about 7-10 days storage	18,000	14,000

Sample Location F2 (liquid digested solids) had low odors compared to the cake samples. Sample Location G (centrifuge prior to conveyance) contained odors much higher than F2 and almost as high as the sample from Sample Location I. The biosolids cake samples were about two orders of magnitude more odorous than the liquid, digested solids sample.

3.1.3 General Observations

The project team noted the following observations:

- The influent wastewater was very septic, showing a relatively high H_2S measurement of 20 ppmv in its headspace. The influent also had a relatively low ORP of 216 millivolts (mV), which reflected the lack of dissolved oxygen and general septicity of the influent wastewater.

- No iron is added to the wastewater influent or within the WWTP itself. The only chemical added to the solids stream is polymer, which is added during the biosolids thickening and dewatering processes.
- The primary sludge (Sample Location C1) had an H_2S concentration of 350 ppmv in its headspace sample, which is the highest H_2S concentration among all the primary sludge samples from the test facilities.
- The digester gas (Sample Location E) had an H_2S concentration of 2,100 ppmv, which is the highest H_2S concentration among all of the digester gas samples from the test facilities.
- Sample Location F2 (liquid digested solids) had low odors compared to the cake samples (from Sample Locations G and I), with an odor difference of two orders of magnitude between the liquid and cake biosolids.
- Sample Location G (centrifuge prior to conveyance) contained odors much higher than F2 and almost as high as the stored biosolids cake sample (Sample Location I).
- WWTP No. 1 was one of two test facilities in which H_2S predominated over all organic sulfur compounds, in terms of total sulfur emissions, in all of the headspace samples analyzed.

3.2 WWTP No. 2

3.2.1 WWTP Description

WWTP No. 2 also served as the "WWTP No. 0" for the trial at the beginning of the field study. The field and laboratory trials on WWTP No. 0 were used for fine tuning the sampling and testing protocol developed for this phase of the project. WWTP No. 2 treats wastewater generated in a large city. The facility influent is approximately 90% domestic wastewater, with the remaining 10% of its flow from industrial wastewater.

On the day of testing, May 30, 2002, the influent flow to the WWTP was 200 mgd. The influent wastewater reaches the WWTP via two gravity sewers and a force main. After the influent passes through six mechanically cleaned bar screens, it is degritted, passed through two flocculation tanks, and sent to five rectangular primary sedimentation tanks with chain and flight mechanisms for sludge collection. Screenings and grit from the headworks are hauled directly to an offsite location for disposal.

Secondary treatment consists of 10, four-pass, pure oxygen rectangular aeration basins, each consisting of four cells in a series. Mixed liquor flows from the aeration basins into 10 rectangular secondary settling basins, with chain and flight mechanisms for sludge collection. Secondary effluent collected in the pump station wet well is conveyed to the WWTP outfall by gravity or five pumps, depending on the water surface level in the wet well. Disinfection is achieved at the WWTP outfall before the effluent is conveyed to a nearby river.

Figure 3-2 depicts the solids flow schematic for WWTP No. 2, along with each sample location. Primary sludge is pumped to a mixing chamber where it is combined with TWAS and sludge from another WWTP located in the city. During the testing, samples were taken out of DAF WAS thickening influent (Sample Location A), combined primary sludge (Sample Location B), and WAS thickening effluent before and after mixing with primary sludge in a mixing chamber (Sample Locations C and D, respectively). The mixing chamber pumps the combined sludge directly to 12 mesophilic anaerobic digesters sequentially, with each digester receiving sludge for nine minutes on each sequence.

Approximately 1.2 mgd of digested sludge flows from the digesters past Sample Location F1 (see Figure 3-2) to one of two sludge holding tanks. The sludge holding tanks provide four additional days of storage in the Biosolids Recycling Center (BRC). Biosolids are dewatered by six low-solids or four high-solids centrifuges. The BRC operates 24 hours a day and produces approximately 160 dry tons per day biosolids cake.

Polymer is added to the centrifuge to aid in dewatering. Following centrifugation, cake is conveyed via screw conveyors to a belt conveyor, which then transfers the dewatered biosolids into trucks through plow discharge. Sample Locations G and I represent low- and high-solids centrifuge dewatered cake samples, respectively. Onsite cake storage varies from one day to weeks.

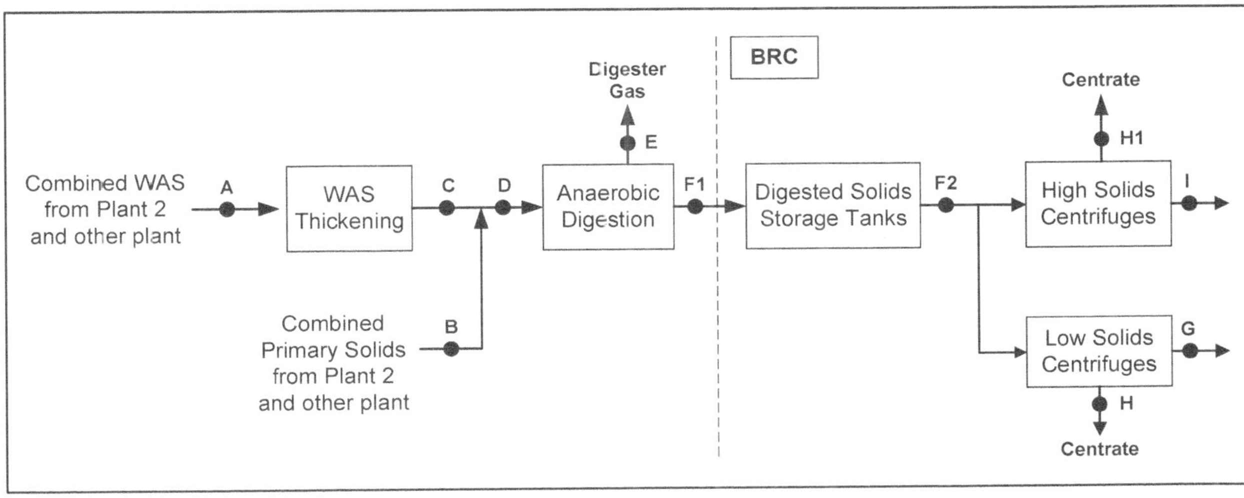

Figure 3-2. Schematic of WWTP No. 2.

3.2.2 General Summary of Results

Table 3-5 presents the results of the WWTP No. 2 field tests on samples collected at Sample Locations B through I. Data from Sample Location A were not available.

Table 3-5. Field Testing Results from WWTP No. 2.

Sample Location	Composite or Grab	Temperature (°C)	PH	ORP (mV)[2]
B Primary Sludge	Composite[1]	21	6.4	-160
C TWAS	Composite[1]	21	6.8	-179
D Combined PS and WAS	Composite[1]	21	6.4	-169
F1 Digested Sludge	Grab	33	7.3	-251
F2 Digested Sludge Post-Storage	Grab	31	7.4	-241
G Low-Solids Cake	Grab	31	--	--
H Low-Solids Centrate	Grab	28	7.6	-12/+48[2]
H1 High-Solids Centrate	Grab	31	7.6	-18/+44[2]
I High-Solids Cake	Grab	33	--	--

Notes:
[1] Three composite samples, separated by one hour (results are average of three composites)
[2] ORP readings with single probe/ion meter

Data from the headspace field testing at WWTP No. 2 are presented in Table 3-6. Duplicates were performed on several H_2S measurements and on one NH_3 measurement. The values in Table 3-6 are the averages of all measurements for each location. Correlations between these results and plant parameters are presented in Chapter 4.0 of this report.

Table 3-6. Field Headspace Testing Results from WWTP No. 2.

Sample Location	H_2S Concentration (ppmv)	NH_3 Concentration (ppmv)
INF Influent	0.003	<0.1
B Primary Sludge	0.49	<0.1
C TWAS	0.013	<0.1
D Combined PS and WAS	0.19	<0.1
E Digester Gas	1.3	<0.1
F1 Digested Sludge	0.01	3
F2 Digested Sludge Post-Storage	0.018	15
G Low-Solids Cake	0.053	5
H Low-Solids Centrate	0.03	19
H1 High-Solids Centrate	0.12	25
I High-Solids Cake	0	4

Table 3-7 presents the results of the odor analysis on samples collected for WWTP No. 2. On June 5, 2002, one week after sampling, St. Croix Sensory measured for odor intensity in headspace gas from all sampling points downstream of the digested sludge storage tanks, with the exception of Sample Locations H and H1 (dewatering centrate samples). Detection and recognition thresholds were measured as part of the olfactometry analysis.

Table 3-7. Odor Evaluation Results from WWTP No. 2.

Sample Location	Detection Threshold (D/T)	Recognition Threshold (R/T)
F2 Digested Solids After Holding Tank	390	230
G Low-Solids Centrifuge Biosolids Cake	6,100	4,300
I High-Solids Centrifuge Biosolids Cake	21,000	13,000
Note: Samples were diluted 1:50 with nitrogen prior to shipment to St. Croix Sensory Laboratory; detection and recognition thresholds of the headspace gases in the sample bottles are 50 times the values reported in this table.		

3.2.3 General Observations

The project team noted the following observations:

- All samples upstream of digestion contained very low (undetectable) NH_3 concentrations in the headspace. It is especially unusual for primary sludge to have undectable NH_3.
- All headspace samples contained fairly low H_2S concentrations, especially in the digester gas (Sample Location E).
- WWTP No. 2 is one of two facilities in the study that contains both high-solids and medium-solids centrifuges in the dewatering process (WWTP No. 10 is the other). This enables a direct comparison of odors and odorous constituents among centrifuge types.
- Sample Location G, which is low-solids centrifuge cake prior to conveyance, contained significantly lower odors than Sample Location I, the high-solids centrifuge cake prior to

conveyance (see Table 3-7). This indicates that the type of dewatering equipment may affect the odor generated from dewatered biosolids.

- Contrary to the olfactometry results, headspace analysis of H_2S and NH_3 indicated little difference between Sample Locations G and I.
- The cake odor results (Samples G and I) are an order of magnitude higher than the liquid sample odor results (Sample F2).

3.3 WWTP No. 3

3.3.1 WWTP Description

This WWTP is located in a large urban area and contains approximately 96% domestic wastewater, with the remaining 4% from a combination of industrial flows from hospitals, metal plating, food processing, and other facilities. WWTP No. 3 has a peak capacity of 250 mgd wet weather flow and treats an average of 68 mgd of dry weather flow. On the day of testing, June 7, 2002, the influent flow was 68 mgd.

The influent wastewater from three major pump stations runs through a series of screens and vortex grit tanks prior to primary and secondary treatment. Screenings from the headworks are hauled to a transfer station offsite, where they are mixed with municipal solid waste (MSW) and disposed in a MSW landfill. WWTP No. 3 is a high-purity oxygen activated sludge plant, with a 30% influent RAS rate. The effluent is chlorinated and dechlorinated before being discharged to a bay through a deep-water outfall.

Figure 3-3 depicts the solids flow schematic for WWTP No. 3, along with the sample locations. The WWTP produces approximately 250,000 gpd of primary sludge, at an average of 4% total solids concentration, through its 11 primary sedimentation tanks (seven were on line during testing). In addition, WWTP No. 3 produces approximately 1.4 mgd of WAS at an average of 0.7 percent total solids concentration. The WAS is subjected to DAF thickening, although PS are not thickened. Six DAF Thickeners (DAFTs) are typically on line at the same time and run 24 hours a day. Polymer is injected into the DAFTs at a rate of 1.5 lb/dry ton.

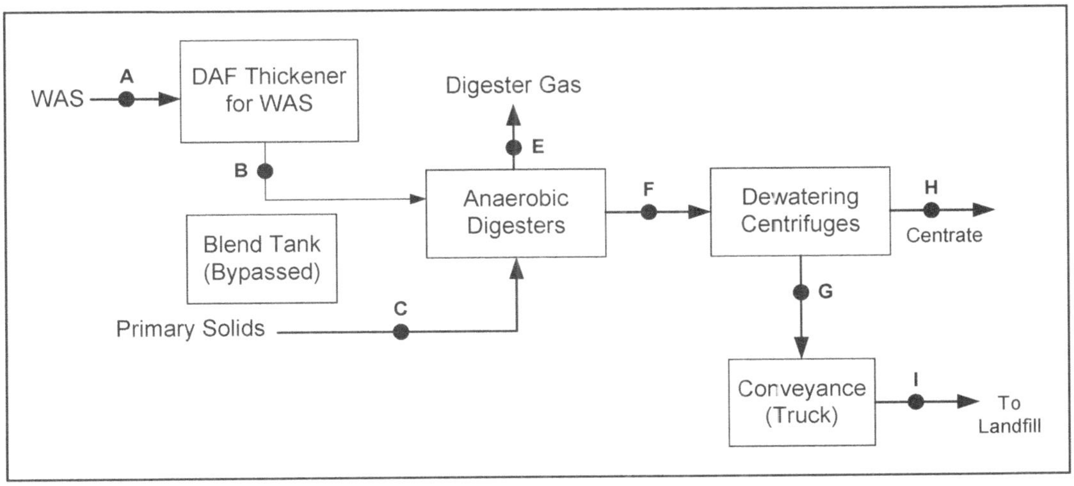

Figure 3-3. Schematic of WWTP No. 3.

WWTP No. 3 contains a blend tank to rapidly mix TWAS with PS prior to digestion, but this tank was not on line during the testing on June 7, 2002, and had not been on line for over two weeks. Instead, both solids trains were piped directly into the digesters, where mixing occurred. Because the two solids streams are mixed in the digesters, there was no Sample Location D for WWTP No. 3 (Figure 3-3) on the day of sampling.

The WWTP has 10 mesophilic anaerobic digesters, all of which contain floating covers, and nine of which were in use on the day of testing. The WWTP has six digesters operational at any given time, with two additional digesters providing secondary storage. Recently WWTP No. 3 increased the temperature in the storage digesters in order to meet U.S. EPA time-temperature requirements. Detention time in the digesters typically ranges between 15 and 20 days. However, the calculation for digester detention time on the day of sampling was 27 days.

Digested biosolids are pumped to six centrifuges for dewatering. The capacities of the centrifuges range from 75–250 gallons per minute. Polymer plus ferric chloride is added to the centrifuges. The resulting biosolids cake is discharged onto a conveyor and transferred to a storage area, where it sits for typically only a few hours, as there is no long-term biosolids storage at the facility. Biosolids from the storage area are loaded directly into a truck bed and taken to a landfill approximately 50 miles away, where they are used as an alternative daily cover. The biosolids can also be taken to a land application site. Sample Location I (Figure 3-3) was taken from a truck bed following up to one day of storage.

3.3.2 General Summary of Results

Table 3-8 presents the results of the field tests on samples collected at Sample Locations A through I. There is no Sample Location D for WWTP No. 3. A detailed presentation of all sampling results is provided in Appendix A of this report, which is available from WERF.

Table 3-8. Field Testing Results from WWTP No. 3.

Sample Location	Composite or Grab	Temperature (°C)	pH	ORP (mV)
A WAS	Composite[1]	22	6.3	-49.2
B TWAS	Composite[1]	21	6.6	-68.3
C Primary Sludge	Composite[1]	21	6.0	-76.7
F Digested Biosolids	Grab	34	7.0	-210
G Dewatered Biosolids	Grab	37	-	-
H Centrate	Grab	35	6.9	-106
I Stored Biosolids Cake	Grab	30	-	-

[1]Three composite samples, separated by one hour (results are average of three composites).

Table 3-9 presents the results of the field headspace analysis on samples collected during WWTP No. 3 testing. Duplicates were performed on several H_2S and NH_3 measurements. Values presented in the table are averages of all measurements.

Table 3-9. Field Headspace Testing Results from WWTP No. 3.

Sample Location	H_2S Concentration (ppmv)	NH_3 Concentration (ppmv)
INF Influent	0.01	1
A WAS	0.45	0
B TWAS	0.04	0
C Primary Sludge	1.5	0
E Digester Gas	11.4[1]	700
F Digested Biosolids	0.14	21
G Dewatered Biosolids	0.03	4
H Centrate	0.03	8
I Stored Biosolids Cake	0.14	3
[1] Measured by Performance Analytical Laboratory.		

Table 3-10 presents the results of the odor analysis on samples collected for WWTP No. 3. On August 6, 2002, one week after sampling, St. Croix Sensory measured for odor intensity in the headspace gas from all sampling points downstream of the anaerobic digesters, with the exception of Sample Location H – dewatering centrate. Detection threshold and recognition threshold were measured.

Table 3-10. Odor Evaluation Results from WWTP No. 3.

Sample Location	Detection Threshold (D/T)	Recognition Threshold (R/T)
F Digested Biosolids	460	270
G Dewatered Biosolids	9,600	7,300
I Stored Biosolids Cake	4,800	4,200
Note: Samples were diluted 1:50 with nitrogen prior to shipment to St. Croix Sensory Laboratory; detection and recognition thresholds of the headspace gases in the sample bottles are 50 times the values reported in this table.		

3.3.3 General Observations

WWTP No. 3 adheres fairly closely to the general process schematic and its solids processing is fairly standard, using DAF thickening, anaerobic digestion, and centrifuge dewatering. The project team noted the following:

- On the day of testing and for more than two weeks prior to testing, the WWTP was not blending PS and TWAS in the blend tank prior to digestion (see Figure 3-4). This could lead to comparisons in the future at WWTP No. 3, if the plant were tested again when the blend tank is back on line.
- Table 3-9 indicates that there was a higher H_2S concentration in the primary sludge sample (Sample Location C) than either of the WAS samples (Sample Locations A or B). This happened in other WWTPs where the two streams were separate.
- Field testing produced low H_2S readings in the digester gas (11 ppmv as analyzed at a nearby laboratory), but high NH_3 concentrations (700 ppmv as measured during testing). In most other WWTPs, the opposite was the case – low NH_3 and higher H_2S readings.
- WWTP No. 3 contained some of the lowest field testing NH_3 headspace concentrations for dewatered cake of all WWTPs in the study.

- Similar to other WWTPs, dewatering of digested biosolids produced much higher odor readings (Table 3-8).
- The minimal storage (up to one day) between Sample Locations G and I decreased odors by approximately half (Table 3-10).

3.4 WWTP No. 4

3.4.1 WWTP Description

WWTP No. 4 is a regional WWTP that treats wastewater generated in five communities. Approximately 34% of the influent flow is a combination of wastewater generated in 12 major industries, most of them related to food processing. WWTP No. 4 is rated for 30 mgd design flow, with a peak wet weather flow design of 60 mgd. On the day of testing, June 18, 2002, the influent flow was approximately 14 mgd.

Preliminary treatment at the WWTP consists of flow measurement, mechanical bar screens, and degritting. Following flow measurement in a parshall flume, flow is diverted to north and south treatment trains. Each train contains two mechanically cleaned bar screens. From there the wastewater is degritted in two circular collectors and passed to four sets of rectangular primary sedimentation tanks with chain and flight mechanisms for sludge collection. Odor has been a major issue at this step of the treatment train, and for this reason, the south train effluent chamber is covered and the headspace gas is sent to the odor scrubbers. Primary sludge flows into a common lift station, where it is pumped to aeration basins by four centrifugal pumps. Screenings and grit from the headworks are hauled directly to a landfill for disposal.

Secondary treatment consists of four four-pass rectangular aeration basins. The first cell is strictly dedicated to RAS and other recycle streams. The remaining three cells can operate either in series or in parallel, with primary effluent being fed to each cell. Mixed liquor then flows into four peripheral fed circular secondary settling basins. The secondary effluent flows into two chlorine contact basins.

Figure 3-4 depicts the solids flow schematic for WWTP No. 4, along with each sample location. The WWTP has a total of eight floating-lid anaerobic digesters arranged in two stages. Four heated digesters are dedicated to the first stage and four non-heated digesters are dedicated to the secondary stage. Digested biosolids are pumped to four sludge lagoons with a combined capacity of 45 million gallons (MG).

Biosolids samples are drawn off the gravity belt WAS thickening influent, the primary sludge, and the trucked waste from industries before they are mixed in the primary digesters. Polymer is added to WAS prior to gravity belt thickening. Digester off-gas is scrubbed with ferrous impregnated wood chips for removal of hydrogen sulfide. The methane is sold to a local gas company.

Sample Locations G and I represent biosolids samples collected from lagoon top and lagoon bottom, respectively (Figure 3-4). Ultimately, biosolids stored in the lagoons for approximately one year are removed by using a dredge and four 6,000-gallon tankers. The solids are then transported to fields for beneficial use, based on crop nutrient needs.

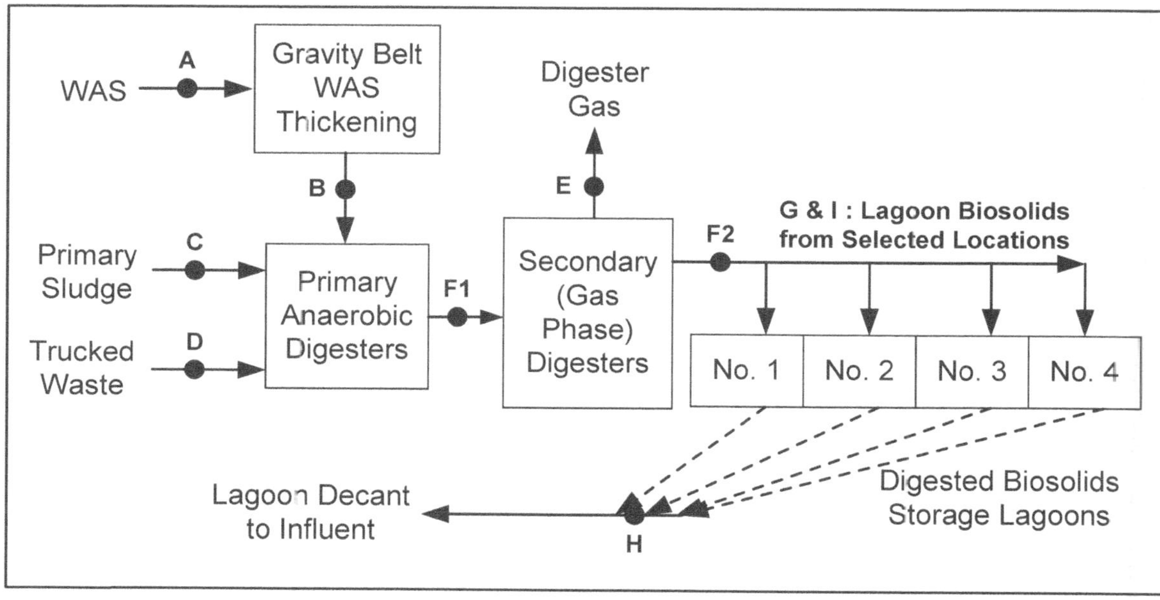

Figure 3-4. Schematic of WWTP No. 4.

3.4.2 General Summary of Results

Data from the field testing at WWTP No. 4 are presented in Table 3-11. Correlations between these results and plant parameters are presented in Chapter 4.0 of this report.

Table 3-11. Field Testing Results from WWTP No. 4.

Sample Location	Composite or Grab	Temperature (°C)	pH	ORP (mV)
INF Influent	Composite	17	6.7	-75
A WAS	Composite[1]	22	6.7	-61
B TWAS	Composite[1]	21	6.5	-65
C Primary Sludge	Composite[1]	21	6.1	-63
D Trucked Waste	Grab	27	5.0	15
F1 Primary Digester	Grab	33	7.0	-213
F2 Digested Biosolids	Grab	25	6.9	-201
G Lagoon Bottom Solids	Grab	22	6.9	-105
H Lagoon Decant	Grab	23	7.6	-184
I Lagoon Top Solids	Grab	24	7.0	-141

[1]Three composite samples, separated by one hour (results are average of three composites).

Table 3-12 presents the results of the field headspace analysis on samples collected during WWTP No. 4 testing. Duplicates were performed on several H_2S and NH_3 measurements. Values presented in the table are averages of all measurements.

Table 3-12. Field Headspace Testing Results from WWTP No. 4.

Sample Location	H_2S Concentration (ppmv)	NH_3 Concentration (ppmv)
INF Influent	1.2	1
A WAS	1.9	Trace
B TWAS	0.05	0
C Primary Sludge	23	Trace
D Trucked Waste	35	0.5
E1 Digester Gas Pre-Scrubber	250	Trace
E2 Digester Gas Post-Scrubber	0.13	0
F1 Primary Digester	6	75
F2 Digested Biosolids	0.5	26
G Lagoon Bottom Solids	3	30
H Lagoon Decant	0.03	42
I Lagoon Top Solids	0.4	15

Table 3-13 presents the results of the odor analysis on samples collected for WWTP No. 4. On June 25, 2002, one week after sampling, St. Croix Sensory measured for odor intensity in the headspace gas from all sampling points downstream of digestion, with the exception of Sample Location H– lagoon supernate sample. Detection and recognition thresholds were measured as part of the olfactometry analysis.

Table 3-13. Odor Evaluation Results from WWTP No. 4.

Sample Location	Detection Threshold (D/T)	Recognition Threshold (R/T)
F2 Digested Sludge	230	120
G Lagoon Top Biosolids Sample	3,700	1,600
I Lagoon Top Biosolids Sample	3,500	2,000
Note: Samples were diluted 1:50 with nitrogen prior to shipment to St. Croix Sensory Laboratory; detection and recognition thresholds of the headspace gases in the sample bottles are 50 times the values reported in this table.		

3.4.3 General Observations

The project team noted the following:

- WWTP No. 4 is located in the most remote area of all facilities in the study, the only one not within 50 miles of a major U.S. city.
- The WWTP has a higher agricultural and food waste influent component than any other facility in the study. Significant quantities of food processing waste are trucked into the facility and deposited into a receiving station, which pumps directly into the primary digester.
- WWTP No. 4 is one of two facilities in the study that uses lagoons for drying digested biosolids (WWTP No. 7 is the other).

- Unlike WWTP No. 7, which also mechanically dewaters the biosolids, lagoon thickening by decanting is the only downstream process used for digested biosolids at WWTP No. 4 (no mechanical dewatering of the biosolids).
- H_2S headspace readings in the field were fairly high in the primary sludge and in the trucked waste.
- High NH_3 headspace measurements were found in the primary digester sample (Sample Location F1).
- The biosolids sample from Sample Location G (solids from the bottom of the lagoon) contained higher concentrations of H_2S and NH_3 in the headspace than the sample for Sample Location I (solids from the top of the lagoon).
- Biosolids from Sample Locations G and I contained similar total odor measurements (Table 3-13), however, indicating that this facility's lagoon biosolids are relatively homogenous in odor characteristics.
- Table 3-13 indicates that the odor results from thickened "lagoon bottom" biosolids are an order of magnitude higher than the liquid "lagoon top" samples.

3.5 WWTP No. 5

3.5.1 WWTP Description

WWTP No. 5 is a small- to medium-sized facility that treats wastewater generated in a mostly suburban, coastal area with a fairly mild climate. WWTP influent consists approximately 90% of domestic and commercial wastewater, with approximately 10% coming from industries comprised mostly of food products (one soft drink manufacturer and several seafood processors). WWTP No. 5 has a nominal capacity of approximately 20 mgd. On the day of testing, June 26, 2002, the influent flow to the WWTP was 13 mgd.

The facility employs a pre-aeration tank for the WWTP influent. Following screening in the headworks, wastewater passes to four circular primary clarifiers. The facility is a conventional activated sludge plant with primary and secondary clarification. Sodium hypochlorite is used for effluent disinfection.

Figure 3-5 depicts the solids flow schematic for WWTP No. 5, along with each sample location. PS are removed from the four primary clarifiers at only about 0.3% solids concentration and pumped to two circular, gravity thickeners, which thicken the PS to about 5% concentration (Sample Location C2).

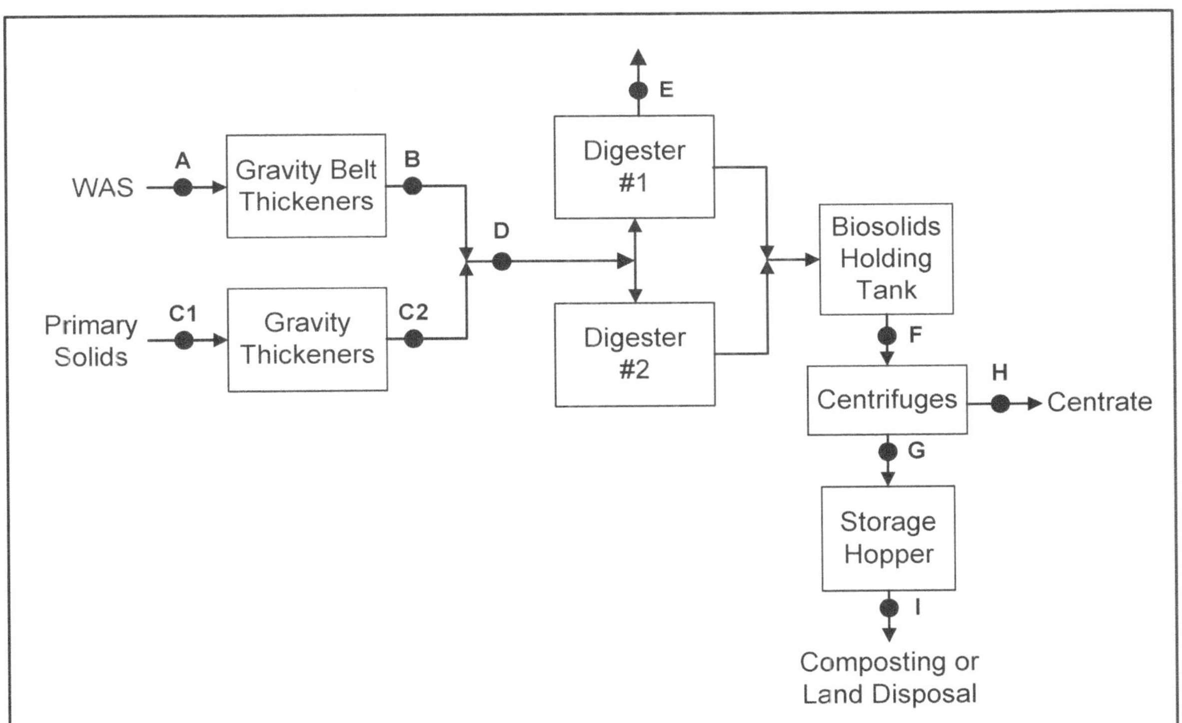

Figure 3-5. Schematic of WWTP No. 5.

WAS is removed from five circular, secondary clarifiers and thickened to 4% total solids concentration by three gravity-belt thickener units. The thickened PS and TWAS are then pumped directly into a single conduit. The conduit splits into two conduits that lead to the two anaerobic digesters (there is no pre-digestion storage), operating in mesophilic mode. The two, single-stage digesters are rated for 20-day detention time at design capacity.

Digested solids flow from the two anaerobic digesters into a holding tank of 0.3-MG capacity. From this post-digestion holding tank, the liquid digested biosolids are pumped to three solid-bowl centrifuges (high-solids type, Alfa Laval DS706 models). The dewatered cake solids, at about 25% dry solids content, drop through a small storage hopper, are held for about five hours, and deposited into trucks. Dewatered biosolids are then transported by truck either to a regional composting facility or less frequently to land application or a land fill.

3.5.2 General Summary of Results

Data from the field testing at WWTP No. 5 are presented in Table 3-14.

Table 3-14. Field Testing Results from WWTP No. 5.

Sample Location	Composite or Grab	Temperature (°C)	pH	ORP (mV)
A WAS	Composite[1]	27	6.3	-98
B TWAS	Composite[1]	27	6.4	-113
C1 Liquid PS	Composite[1]	26	6.7	-228
C2 Thickened PS	Composite[1]	26	5.9	-265
D Combined PS and WAS	Grab	27	6.4	-180
F Digested Biosolids	Grab	34	7.0	-240
G Centrifuged Biosolids Cake	Grab	27	7.8	-181
H Centrate	Grab	34	7.6	-112
I Stored Biosolids Cake	Grab	28	8.0	-140

[1]Three composite samples, separated by one hour (results are average of three composites)

Table 3-15 presents the results of the field headspace analysis on samples collected during WWTP No. 5 testing. Duplicates were performed on several H_2S and NH_3 measurements. Values presented in the table are averages of all measurements.

Table 3-15. Field Headspace Testing Results from WWTP No. 5.

Sample Location	H_2S Concentration (ppmv)	NH_3 Concentration (ppmv)
INF Influent	105	Trace
A WAS	0.02	0
B TWAS	0.01	Trace
C1 Liquid PS	130	0.2
C2 Thickened PS	14	0
D Combined PS and WAS	0.03	0
E Digester Gas	15	0.25
F Digested Biosolids	0.02	53
G Centrifuged Biosolids Cake	1.2	10
H Centrate	0.17	40
I Stored Biosolids Cake	0.3	10

Table 3-16 presents the results of the odor analysis on samples collected for WWTP No. 5. On July 3, 2002, one week after sampling, St. Croix Sensory measured for odor intensity in the headspace gas from all sampling points downstream of digestion, with the exception of Sample Location H – dewatering centrate samples. Detection and recognition thresholds were measured as part of the olfactometry analysis.

Table 3-16. Odor Evaluation Results from WWTP No. 5.

Sample Location	Detection Threshold (D/T)	Recognition Threshold (R/T)
F Digested Biosolids	270	140
G High-Solids Centrifuge Biosolids Cake	6,100	3,500
I High-Solids Centrifuge Cake after about 2 days storage	7,400	4,300
Note: Samples were diluted 1:50 with nitrogen prior to shipment to St. Croix Sensory Laboratory; detection and recognition thresholds of the headspace gases in the sample bottles are 50 times the values reported in this table.		

3.5.3 General Observations

The project team noted the following observations:

- Similar to WWTP No. 1, the influent wastewater of WWTP No. 5 was very septic, with a H_2S concentration of 105 ppm, the highest H_2S measurement in the headspace of all influent samples tested.
- The primary sludge headspace sample had a relatively high H_2S concentration of 130 ppm.
- Headspace samples for WAS and TWAS had very low H_2S concentrations, possibly due primarily to the addition of ferrous chloride to the aeration basin mixed liquor before it enters the secondary clarifiers.
- The digester gas had a relatively low H_2S concentration of 15 ppm. Again, this is likely due to the ferrous chloride addition upstream of the digesters.
- The headspace odor samples of dewatered biosolids were some of the lowest among facilities with high-solids centrifuges, which may be due in part to the addition of ferrous chloride to the mixed liquor in the facility.
- The two-day storage of the biosolids cake increased odors slightly (between Sample Locations G and I).

3.6 WWTP No. 6

3.6.1 WWTP Description

WWTP No. 6 treats wastewater from a portion of a large city. The WWTP is located in a somewhat isolated industrial area and is bordered by a large river to the east. The nearest residential community is about 1.5 miles away. The facility influent is approximately 90% domestic wastewater, with the remaining 10% of its flow from a combination of metals and organic material from various nearby industries.

WWTP No. 6 treats an average of 60 mgd dry weather flow in its activated sludge plant. On the day of testing, July 9, 2002, the influent flow was 55 mgd. Before primary and secondary treatment, the influent wastewater from three interceptors that extend through the city is run through a series of nine screens and five grit chambers. Screenings from the headworks are hauled directly to the landfill located across the street from the WWTP. Four primary settling tanks provide treatment, and eight aeration and eight final settling basins complete the secondary process train. Secondary effluent is conveyed through 12 pumps into two tertiary treatment polishing ponds, which provide two days residence time. The pond effluent is chlorinated and then discharged to the nearby river.

Figure 3-6 depicts the solids flow schematic for WWTP No. 6, along with the sample locations. Solids are drawn off from the primary settling tanks and combined with WAS from the final settling basins in a sludge mixing chamber, which is a small tank providing minimal storage time. The combined sludge is pumped directly to three gravity thickeners, bypassing the holding tanks.

The gravity sludge thickener underflow is conveyed out of the center of the tank. The sludge is split between the anaerobic digesters and a second stage of thickening (Figure 3-6). Prior to the second stage of thickening, two GBT holding tanks provide about one day of sludge storage. GBTs perform the second stage of thickening, typically for 60-75% of the sludge; 67% of the sludge received this second stage of thickening on the date of testing. Polymer is injected into the thickeners. All sludge is recombined prior to entering the five anaerobic digesters (Sample Location D), which provide 21-26 days residence time. There are six mesophilic anaerobic digesters at WWTP No. 6, all with floating covers, but one has been out of service since 1999.

Approximately 0.25 mgd sludge flows from the digesters past Sample Location F (Figure 3-6). All digested sludge is conveyed to one of two sludge holding tanks, which provide another day of storage. The one centrifuge at the WWTP, which operates 24 hours a day and produces a 20% solids biosolids cake, dewaters all digested biosolids. As a backup to the centrifuge, WWTP No. 6 has a belt filter press available. Polymer is added to the centrifuge to aid in dewatering. Following centrifugation, cake drops into a hopper and then is conveyed through high-pressure sludge pumps approximately 200′ into a storage pile. This conveyance system (resulting in Sample Location I) is completely enclosed, and polymer is injected prior to pumping. Front-end loaders pick up the cake from the pile, mix the sludge with fly ash, and haul it across the street to a landfill, where it is used as a daily cover.

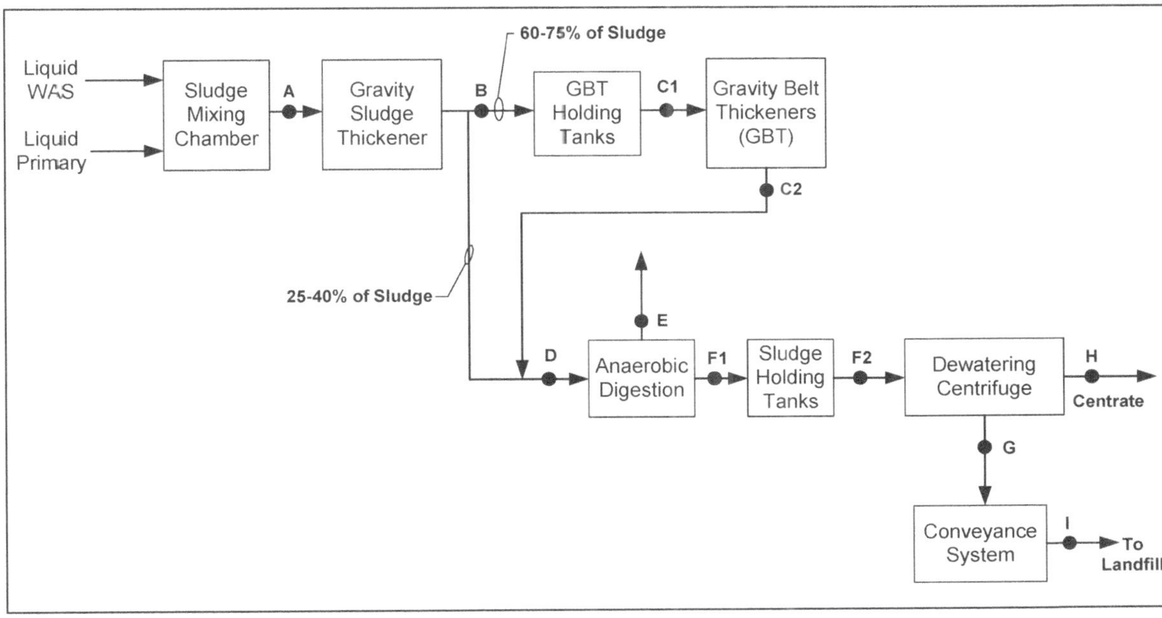

Figure 3-6. Schematic of WWTP No. 6.

3.6.2 General Summary of Results

Table 3-17 presents the results of the field tests on samples collected in Sample Locations A through I for WWTP No. 6, with the exception of downstream of the post-digestion holding tank, where field readings of temperature, pH, and ORP were not taken.

Table 3-17. Field Testing Results from WWTP No. 6.

Sample Location	Composite or Grab	Temperature (°C)	pH	ORP (mV)
A Combined Sludge (CS)	Composite[1]	24	6.7	-155
B Thickened CS (TCS)	Composite[1]	26	6.0	-191
C1 TCS After Holding Tank	Composite[1]	26	5.9	-167
C2 Gravity-Belt TCS	Composite[2]	26	5.9	-195
D Digester Feed Sludge	Composite[1]	24	6.0	-183
F1 Digested Biosolids	Grab	32	7.0	-268
F2 Digested Biosolids After Holding Tank	Grab	33	7.1	-232
G Fresh Biosolids Cake[3]	Grab	-	-	-
H Centrate[3]	Grab	-	-	-
I Stored Biosolids Cake[3]	Grab	-	-	-

Notes:
[1]Three composite samples, separated by one hour (results are average of three composites).
[2]Two composite samples, separated by one hour (results are average of three composites).
[3]Cake and centrate samples were taken by WWTP staff – no field readings taken.

Data from the headspace field testing at WWTP No. 6 are presented in Table 3-18. Duplicates were performed on several H_2S measurements and on one NH_3 measurement. The table presents the averages of all field measurements.

Table 3-18. Field Headspace Testing Results from WWTP No. 6.

Sample Location	H_2S Concentration (ppmv)	NH_3 Concentration (ppmv)
INF Influent	0.32	1.5
A Combined Sludge (CS)	3.5	<1.0
B Thickened CS (TCS)	0.23	2.5
C1 TCS After Holding Tank	4.8	<1.0
C2 Gravity-Belt TCS	2.3	<1.0
D Digester Feed Sludge	5.3	<1.0
E Digester Gas	0.31	6
F1 Digested Biosolids	0.17	28
F2 Digested Biosolids After Holding Tank	0.06	65
G Fresh Biosolids Cake	0.19	46
H Centrate	0.03	62
I Stored Biosolids Cake	0.16	51

Table 3-19 presents the results of the odor analysis on samples collected for WWTP No. 6. On July 16, 2002, one week after sampling, St. Croix Sensory measured for odor intensity in the headspace gas from all sampling points downstream of digestion, with the exception of Sample Location H – dewatering centrate. Detection threshold and recognition threshold were measured.

Table 3-19. Odor Evaluation Results from WWTP No. 6.

Sample Location	Detection Threshold (D/T)	Recognition Threshold (R/T)
F1 Digested Biosolids	95	70
F2 Digested Biosolids After Holding Tank	120	75
G Fresh Biosolids Cake	5,100	3,100
I Stored Biosolids Cake	2,900	1,700
Note: Samples were diluted 1:50 with nitrogen prior to shipment to St. Croix Sensory Laboratory; detection and recognition thresholds of the headspace gases in the sample bottles are 50 times the values reported in this table.		

3.6.3 General Observations

WWTP No. 6 is a medium-sized WWTP in the study. Like several other WWTPs in the study, WWTP No. 6 utilizes biosolids holding tanks post-digestion and uses centrifugation for dewatering. The project team noted the following observations:

- WWTP No. 6 is one of two WWTPs that mixes PS and WAS prior to thickening (WWTP No. 7 is the other facility). All other WWTPs in the study mix their PS and WAS at the digestion step.
- The WWTP is the only facility in the study that provides two stages of thickening for a portion of its solids train (Figure 3-6). On the date of testing, 67% of the gravity-thickened sludge passed through a separate holding tank and then was thickened a second time by GBTs.

- Dewatered biosolids cake is conveyed through a 200′ pipeline by means of a high-pressure sludge cake pump.
- The first stage of thickening of the combined solids (in the gravity sludge thickener) produced a decrease in the field-tested H_2S headspace concentration and an increase in NH_3 headspace concentration. A similar effect was noted in WWTP No. 7 for the H_2S readings, though NH_3 concentrations remained approximately the same.
- Table 3-18 indicates that there was a large increase in field headspace H_2S concentration following the sludge being detained the GBT holding tank.
- Field testing produced very low H_2S readings in the digester gas (less than 1 ppmv) and fairly low NH_3 concentrations.
- WWTP No. 6 contained high field testing NH_3 headspace concentrations for points downstream of digestion, including a more than 100% increase following the digested sludge holding tank (between Sample Locations F1 and F2).
- Similar to other WWTPs, dewatering of digested biosolids produced an order of magnitude higher odor readings (Table 3-19). Additionally, Table 3-19 indicates that storage of the digested biosolids increases odor slightly at WWTP No. 6.
- Sample Location G, where the cake is sampled prior to conveyance through the sludge pumps, contained more than 75% greater detection threshold than that of Sample Location I, where the cake was taken from the biosolids pile following conveyance.

3.7 WWTP No. 7

3.7.1 WWTP Description

WWTP No. 7 predominantly treats municipal wastewater from suburban communities. Approximately 5% of the influent flow is from industrial discharges, such as plating shops and circuit board manufacturers. The WWTP is located alongside a large bay and the nearest residential community is about 0.5 miles away.

WWTP No. 7 treats an average of 17 mgd dry weather flow. On the day of testing, July 19, 2002, the influent flow was 17 mgd. In its process scheme, WWTP No. 7 has primary sedimentation basins followed by fixed-film reactors and an aerated activated sludge system. Following secondary clarification, liquids are disinfected with hypochlorite and run through dual media filters. Backwash from these filters is returned to the WWTP headworks. The final liquid treatment steps involve chlorination and then dechlorination with bisulfite. Effluent is either discharged to the nearby bay or used for recycled water.

Solids processed at the facility include materials from primary and secondary clarification, as depicted in Figure 3-7. Approximately 1.5 mgd of PS are run through a grit removal system and then mixed with secondary solids (about 0.4 mgd) in a blend tank. All of the blended solids are treated in two gravity thickeners, with a detention time of approximately 15 hours. Thickened solids are fed into one mesophilic anaerobic digester, which has a fixed cover and a 14-day residence time.

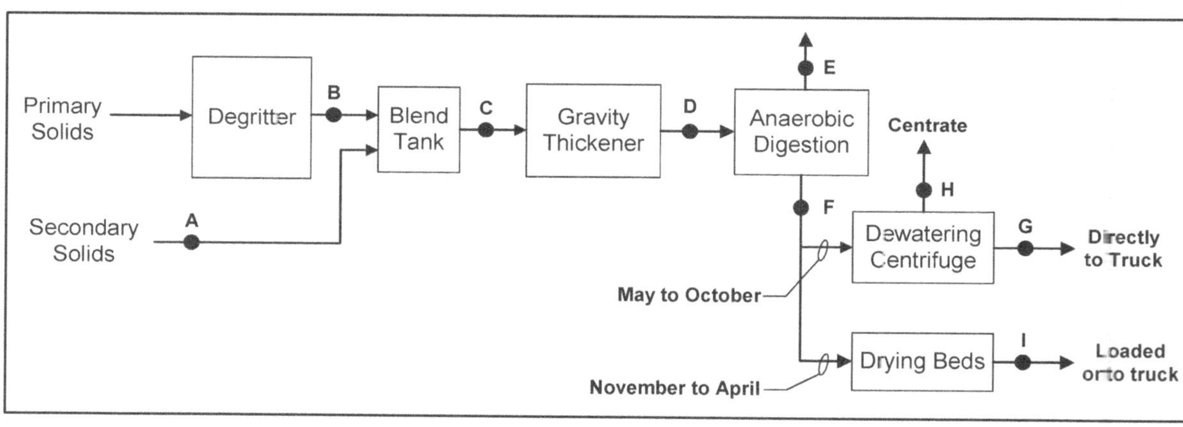

Figure 3-7. Schematic of WWTP No. 7.

Digested solids are treated in one of two ways, depending on the time of year. From November to April of any given year, digested solids are pumped to drying beds on the periphery of the WWTP grounds. The solids are allowed to dry through the summer and then loaded onto trucks by front loaders for removal and disposal in a landfill or for agricultural reuse. Dewatered sludge from the drying beds averages around 50% solids. Sample Location I (Figure 3-7) is representative of biosolids that have been stored through part of the summer, as it was sampled directly from the drying bed.

From May to October, digested solids are pumped to one centrifuge. Polymer is added to this dewatering process. Cake from the centrifuge averages approximately 27% solids. After dewatering, the cake is carried immediately up a conveyor and dumped directly into one of four trailers in the loading bay. Once a trailer is full, it may remain parked onsite for several days before the solids are hauled away. Biosolids are either disposed of at a landfill or employed for agricultural reuse. The sample from Sample Location G was taken directly out of the centrifuge (no storage).

3.7.2 General Summary of Results

Table 3-20 presents the results of the field tests on samples collected in Sample Locations A through I for WWTP No. 7.

Table 3-20. Field Testing Results from WWTP No. 7.

Sample Location	Composite or Grab	Temperature (°C)	pH	ORP (mV)
A Secondary Solids	Composite[1]	24	7.1	-82
B Degritted PS	Composite[1]	23	7.4	-114
C Blended Solids	Composite[1]	24	7.2	-115
D Gravity Thickened Solids	Composite[1]	24	5.6	-204
F Digested Biosolids	Grab	37	7.0	-307
G Centrifuge Biosolids Cake	Grab	36	8.0	-
H Centrate2	Grab	34	7.8	-
I Drying Bed Biosolids Cake	Grab	21	7.4	-

[1]Three composite samples, separated by one hour (results are average of three composites).

Data from the headspace field testing at WWTP No. 7 are presented in Table 3-21. Duplicates were performed on several H_2S and NH_3 measurements. Values shown in the table are representative of averages of all measurements.

Table 3-21. Field Headspace Testing Results from WWTP No. 7.

Sample Location	H_2S Concentration (ppmv)	NH_3 Concentration (ppmv)
INF Influent	0.07	1
A Secondary Solids	2.4	0
B Degritted PS	22.5	2
C Blended Solids	16.0	1.5
D Gravity Thickened Solids	1.7	1.25
E Digester Gas	2,000	0
F Digested Biosolids	5.6	17
G Centrifuged Biosolids Cake	0.9	12
H Centrate	0.05	18.5
I Drying Bed Biosolids Cake	0.05	17

Table 3-22 presents the results of the odor analysis on samples collected for WWTP No. 7. On July 26, 2002, one week after sampling, St. Croix Sensory measured for odor intensity in the headspace gas from all sampling points downstream of digestion, with the exception of Sample Location H – dewatering centrate.

Table 3-22. Odor Evaluation Results from WWTP No. 7.

Sample Location	Detection Threshold (D/T)	Recognition Threshold (R/T)
F Digested Biosolids	1,300	830
G Centrifuge Biosolids Cake	19,000	14,000
I Drying Bed Biosolids Cake	1,900	1,400
Note: Samples were diluted 1:50 with nitrogen prior to shipment to St. Croix Sensory Laboratory; detection and recognition thresholds of the headspace gases in the sample bottles are 50 times the values reported in this table.		

3.7.3 General Observations

WWTP No. 7 was the smallest WWTP in the study in terms of average daily wastewater flow, at 17 mgd. The project team noted the following:

- WWTP No. 7 uses a degritter on its PS prior to conveyance into the blend tank. After consideration, the project team opted not to sample prior to the degritter.
- The WWTP is one of two WWTPs that mixes PS and WAS prior to thickening (WWTP No. 6 is the other one).
- WWTP No. 7 is the only WWTP to use two different types of dewatering, dependent upon the season (centrifuge in the summer and drying beds in the winter).
- Degritted PS samples contained relatively high field-tested H_2S headspace concentrations at 27 ppmv (exceeded only by WWTPs No. 1 and No. 5).
- Thickening the combined solids produced a decrease in the field-tested H_2S headspace concentration similar to WWTP No. 6.
- This WWTP had very high H_2S readings in the digester gas (over 2,000 ppmv, very similar to WWTP No. 1) and undetectable NH_3.
- Similar to other WWTPs, cake odor results (Samples G and I) were higher than the liquid at Sample Location F (sludge directly out of the digester).
- Sample Location G (cake directly out of the centrifuge) contained by far the highest odors of the three sampling points tested, with exactly an order of magnitude higher odor than cake from Sample Location I (cake taken from the drying bed).

3.8 WWTP No. 8

3.8.1 WWTP Description

WWTP No. 8 is a large facility that treats predominantly municipal wastewater from urban and suburban communities. Approximately 15% of the influent flow is industrial, made up of wastewater from food services, metal shops, and other industries. The WWTP sends its treated wastewater through an ocean outfall into the Pacific Ocean. The nearest residence to the WWTP is 100′ away.

On the day of testing, July 30, 2002, the influent flow at WWTP No. 8 was approximately 350 mgd. WWTP No. 8 employs fully primary and secondary treatment of its wastewater. The influent first passes through the headworks, which include bar screen and degritters. Primary treatment is accomplished with 20 primary sedimentation basins, which produce approximately 3 mgd of sludge. Secondary treatment is performed in 12 activated sludge reactors, which inject 96% pure oxygen. Following secondary clarification, most of the effluent is discharged through a five-mile ocean outfall. About 6% of the WWTP effluent is pumped to a water reclamation plant to produced recycled water.

Solids processed at the facility include materials from primary and secondary clarification, as depicted in Figure 3-8. The WWTP thickens approximately 0.8 mgd of WAS through seven centrifuges, a process that includes dosing with polymer. WWTP No. 8 is the only facility in the study to use centrifuges for sludge thickening. TWAS and PS are mixed directly in the digesters, as there is no blend tank prior to digestion. Prior to being pumped into the digester, Fe salts (ferrous chloride) are injected into the primary sludge.

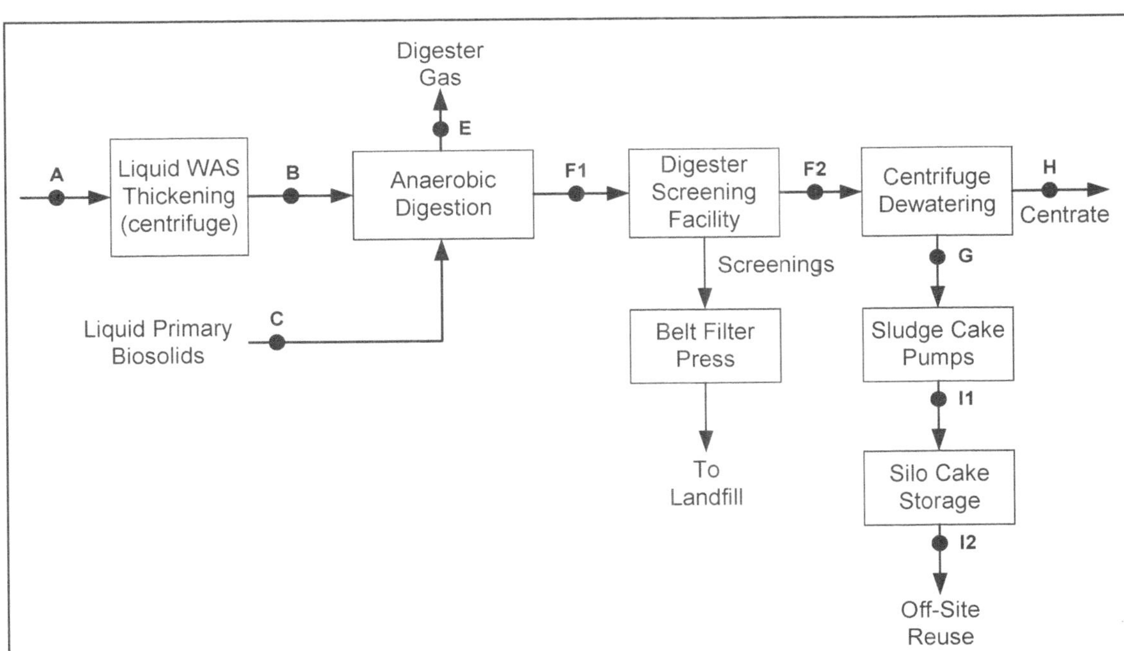

Figure 3-8. Schematic of WWTP No. 8.

WWTP No. 8 is the only WWTP in the study to employ thermophilic digestion of sludge. During the date of sampling, of the 22 digesters in service, 16 were thermophilic, running at an average temperature of 37°C (mesophilic digesters were running at an average temperature of 22°C). On the date of sampling, the actual sludge flow into the thermophilic digesters was 4.18 mgd, including both primary sludge and WAS. The flow into the mesophilic digesters was only 0.35 mgd on the date of sampling and only included PS. Therefore, overall on the date of sampling, WWTP No. 8 was running at approximately 93% thermophilic digestion.

WWTP No. 8 is also the only facility in the study to include a digested sludge screening facility (Figure 3-8), which removes small particles such as hair from the sludge. The screenings themselves are processed in a belt filter press, but were not tested in the study, as they make up approximately 1% of the total dewatered sludge from the facility. Sample Location F2 (post-screening) is the first place that digested sludge is exposed to the air.

Following screening, digested solids are pumped to four centrifuges for dewatering, with the aid of polymer addition. On the date of sampling, 4.0 mgd was sent through the centrifuges. Cake from the centrifuge averages about 32.5% solids. After dewatering, the cake is conveyed through high-pressure pumps (3,000 pounds per square inch [psi]) into eight storage silos, which at any given time hold a total of about 300 tons of sludge. Sample Location I2 represents approximately one day of storage in the silos. Trucks haul the finished biosolids to offsite reuse points between 10 p.m. and 9 a.m. daily.

3.8.2 General Summary of Results

Table 3-23 presents the results of the field tests on samples collected at Sample Locations A through I2 (there is no Sample Location D) for WWTP No. 8.

Table 3-23. Field Testing Results from WWTP No. 8.

Sample Location	Composite or Grab	Temperature (°C)	pH	ORP (mV)
A WAS	Composite[1]	28	6.7	-155
B TWAS	Composite[1]	27	6.8	-174
C Liquid PS	Composite[1]	28	6.1	-116
F1 Digested Sludge	Grab	48	7.4	-256
F2 Digested Sludge Post-Screening	Grab	44	7.6	-125
G Centrifuge Biosolids Cake	Grab	49	--	--
H Centrate	Grab	49	7.9	-44
I1 Cake Post-Conveyance	Grab	51	--	--
I2 Cake Post-Storage	Grab	36	--	--

[1]Three composite samples, separated by one hour (results are average of three composites).

Table 3-24 presents the results of the field headspace analysis on samples collected during WWTP No. 8 testing. Duplicates were performed on several H_2S and NH_3 measurements. Values presented in Table 3-24 are averages of all measurements taken.

Table 3-24. Field Headspace Testing Results from WWTP No. 8.

Sample Location	H2S Concentration (ppmv)	NH3 Concentration (ppmv)
INF Influent	0.10	0.5
A WAS	0.38	1.5
B TWAS	1.8	1.5
C Liquid PS	2.[illegible]	0
E Digester Gas	200	0
F1 Digested Sludge	0.01	35
F2 Digested Sludge Post-Screening	0.02	38
G Centrifuge Biosolids Cake	0.01	53
H Centrate	0.01	50
I1 Cake Post-Conveyance	0.02	48
I2 Cake Post-Storage	0.36	198

Table 3-25 presents the results of the odor analysis on samples collected for WWTP No. 8. On August 6, 2002, one week after sampling, St. Croix Sensory measured for odor intensity in the headspace gas from all sampling points downstream of the digested sludge screening facility, with the exception of Sample Location H – dewatering centrate.

Table 3-25. Odor Evaluation Results from WWTP No. 8.

Sample Location	Detection Threshold (D/T)	Recognition Threshold (R/T)
F2 Digested Sludge Post-Screening	120	65
G Centrifuge Biosolids Cake	9,100	5,000
I1 Cake Post-Conveyance	2,500	1,400
I2 Cake Post-Storage	8,900	6,100
Note: Samples were diluted 1:50 with nitrogen prior to shipment to St. Croix Sensory Laboratory; detection and recognition thresholds of the headspace gases in the sample bottles are 50 times the values reported in this table.		

3.8.3 General Observations

WWTP No. 8 is one of the more complex WWTPs in the study, and has several unique aspects. The project team noted the following:

- The WWTP is the only WWTP that uses thermophilic digestion. On the date of sampling, the WWTP was running at approximately 93% thermophilic digestion, which is reflected in the high temperature readings in Table 3-23 downstream of digestion.
- WWTP No. 8 is also the only WWTP in the study that employs physical treatment of the sludge following digestion, as the digester screening facility removes small items such as hair from the digested biosolids. These screenings are processed in a belt filter press.
- WWTP No. 8 uses high-pressure sludge cake pumps to convey dewatered cake to silos for temporary storage.

- WWTP No. 8 is the only WWTP in the study that uses centrifuges for WAS thickening. As depicted in Table 3-24, the H_2S concentration in the headspace (field tested) increased by an order of magnitude following centrifugation thickening.
- H_2S and NH_3 field headspace measurements were very similar for all samples downstream of digestion, with the exception of Sample Location I2, which was considerably higher (Table 3-24).
- There was a dramatic difference in the field headspace testing for H_2S, and in particular, NH_3 concentrations between Sample Locations I1 and I2 after approximately one day of storage in the silos.
- Similar to other WWTPs, dewatered digested biosolids produced significantly high odor readings (Table 3-25).
- Similar to the field headspace analysis results, Sample Location I2 contained higher odor than cake from Sample Location I1. However, odor decreased following the high-pressure cake conveyance (between Sample Locations G and I1), which is similar to the odor results from WWTP No. 6.

3.9 WWTP No. 9

3.9.1 WWTP Description

WWTP No. 9 predominantly treats municipal wastewater from suburban communities, serving a population of approximately 100,000 people. Approximately 9% of the influent flow is from commercial or industrial discharges, such as a prison laundromat and an amusement park. WWTP No. 9 treats an average of approximately 11 mgd dry weather flow, making it one of the smallest in the study. On the day of testing, August 7, 2002, the influent flow was 13 mgd.

WWTP No. 9 treats wastewater from its collection system as well as effluent and WAS from a nearby water reclamation facility. The combined influent passes through communitors and aerated grit chambers on its way to the primary sedimentation tanks. Fe salts (Fe sulfate) are added to the influent ahead of the communitors in the process train. WWTP No. 9 wastewater undergoes secondary treatment in aerated activated sludge tanks and settling tanks, with a RAS rate of about 45%. Following secondary clarification, liquids pass through dual-media pressure filters and are disinfected inside a chlorine contact tank. Effluent is used for recycled water.

Solids processed at the facility include materials from primary and secondary clarification, as depicted in Figure 3-9. Approximately 95,000 gpd of PS are generated out of primary treatment. These solids are not thickened but sent directly into the digesters. The WWTP No. 9 WAS flowrate is about 0.27 mgd, including the WAS flow from the adjacent WWTP. WAS samples from the WWTP No. 9 secondary treatment and the adjacent facility were mixed at a flow-weighted ratio to best represent the WAS flow entering the sludge thickening stage 9 (Sample A).

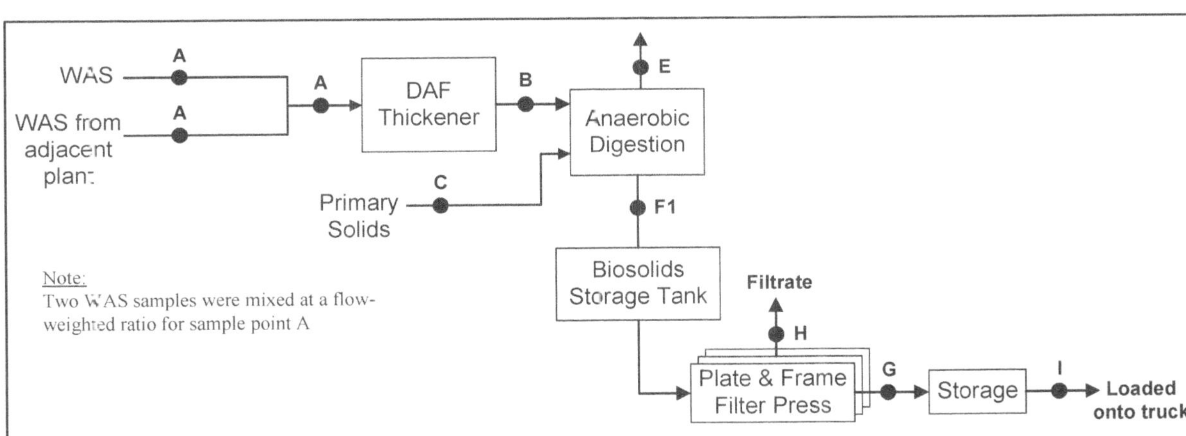

Figure 3-9. Schematic of WWTP No. 9.

All WAS is thickened before digestion in one DAFT. Polymer is dosed into the DAFT, which runs 24 hours a day and provides a 2.5-day detention time for the sludge. After thickening, the WAS is distributed directly into the four digesters, at which time it mixes with PS. Because WAS mixes with PS only in the digesters, the project team determined that there is no Sample Location D at WWTP No. 9.

Thickened solids are distributed into four mesophilic anaerobic digesters, all of which have fixed covers and provide a 19-day residence time for digestion. Because the WAS and PS are equally distributed between the four digesters at WWTP No. 9, the onsite sampling team created a spatial composite from all four digesters in order to produce sample F1 (Figure 3-9). The team took an equal amount of sludge from each of the four digesters and mixed the bucket thoroughly prior to testing and preparation of the bottles.

All digested sludge is conveyed to a sludge holding tank immediately downstream of the digesters. This 230,000-gallon tank provides about one day of storage prior to dewatering and produces Sample Location F2, which is sampled directly from the storage tank. All digested sludge is pumped to one of two plate and frame filter presses for dewatering. At the time of sampling (and typically at WWTP No. 9), only one filter presses was on line. WWTP No. 9 uses the filter press 16 hours a day, Monday through Friday, and doses polymer into the unit. Filtrate is sent to an equalization tanks, and the dewatered cake drops from the belt filter presses directly into trucks.

The trucks typically haul the biosolids away within a few hours either to a landfill or to land applications; however, sometimes a truck can sit onsite over the weekend with cake inside until it is hauled away on Monday morning. Sample Location I was created to simulate this situation, as a bucket of cake was collected on August 5, 2002 and bottled (after 2.5 days of storage) on August 7, 2002 during the testing at WWTP No. 9.

3.9.2 General Summary of Results

Table 3-26 presents the results of the field tests on samples collected in Sample Locations A through I for WWTP No. 9 (there is no Sample Location D).

Table 3-26. Field Testing Results from WWTP No. 9.

Sample Location	Composite or Grab	Temperature (°C)	pH	ORP (mV)
A WAS	Composite[1]	28	7.1	56
B TWAS	Composite[1]	28	7.0	-
C Liquid PS	Composite[1]	27	5.8	-5
F1 Digested Sludge	Composite[2]	34	7.2	-39
F2 Digested Sludge Post-Storage	Grab[3]	34	7.2	33
G Plate & Frame Filter Press Cake	Grab	33	--	--
H Filter Press Filtrate	Grab	33	7.1	59
I 2-Day Stored Cake	Grab	26	--	--

Notes:
[1]Three composite samples, separated by one hour (results are average of three composites).
[2]The F1 sample was a spatial composite, taken from each of the four digesters at approximately the same time.
[3]The F2 sample was taken directly from the biosolids storage tank.

Data from the headspace field testing at WWTP No. 9 are presented in Table 3-27. Duplicates were performed for Sample Location B. Values presented in Table 3-27 are averages of all measurements taken.

Table 3-27. Field Headspace Testing Results from WWTP No. 9.

Sample Location	H2S Concentration (ppmv)	NH3 Concentration (ppmv)
INF Influent	0.11	0.5
A WAS	0.06	0
B TWAS	0.22	0
C Liquid PS	5.2	0.5
E Digester Gas	90	2
F1 Digested Sludge	0.15	40
F2 Digested Sludge Post-Storage	0.12	50
G Plate & Frame Filter Press Cake	2.1	10
H Filter Press Filtrate	0.09	19
I Cake Stored for Two Days	5.0	15

Table 3-28 presents the results of the odor analysis on samples collected for WWTP No. 9. On August 14, 2002, one week after sampling, St. Croix Sensory measured for odor intensity in the headspace gas from all sampling points downstream of digestion, with the exception of Sample Location H – dewatering filtrate.

Table 3-28. Odor Evaluation Results from WWTP No. 9.

Sample Location	Detection Threshold (D/T)	Recognition Threshold (R/T)
F1 Digested Sludge	2,500	1,400
F2 Digested Sludge Post-Storage	95	70
G Plate & Frame Filter Press Cake	1,700	1,100
I Cake Stored for Two Days	2,200	1,300
Note: Samples were diluted 1:50 with nitrogen prior to shipment to St. Croix Sensory Laboratory; detection and recognition thresholds of the headspace gases in the sample bottles are 50 times the values reported in this table.		

3.9.3 General Observations

WWTP No. 9 is one of the smaller WWTPs in the study, similar in capacity to WWTPs No. 6 and No. 7. The following were noted by the project team:

- WWTP No. 9 is the only facility in the study that regularly mixes WAS from another facility with its own WAS.
- Due to limitations of sample locations within the WWTP, liquid samples were taken directly out of the digesters (Sample Location F1) and biosolids storage tank (Sample Location F2).
- WWTP No. 9 is also the only facility in the study that uses a plate and frame filter press for dewatering digested sludge.

- Thickening the WAS produced only a slight increase in the field-tested H_2S headspace concentration (from an average of 0.06 ppmv for Sample Location A to an average 0.22 ppmv for Sample Location B).
- The digesters at this plant did not get nearly as anaerobic as those at the other WWTPs (ORP of -39 mV vs < -150 mV at all other WWTPs).
- Field-tested headspace NH_3 concentrations were higher in the liquid digested sludge (Sample Locations F1 and F2) than in the dewatered cake (Sample Locations G and I).
- Storage of the sludge for 2.5 days (Sample Locations G versus I) produced an increase in both H_2S and NH_3 field headspace concentrations.
- Three of the four sampling points measured for total odor revealed fairly similar results, but the odor readings from Sample Location F2 (digested sludge following about one day of storage) were much lower.

Table 3-28 indicates a slight increase in cake odor during the 2.5-day storage period, though not as great of an increase as was measured in the cake following dewatering.

3.10 WWTP No. 10

3.10.1 WWTP Description

WWTP No. 10 is a large facility that predominantly treats municipal wastewater from urban and suburban communities, with approximately 20% of the influent flow from industries, including refineries, metal platers, and other industrial operations. It treats about 350 mgd and serves approximately 3.5 million people. On the date of testing (August 15, 2002) the WWTP influent was 345 mgd. WWTP No. 10 is surrounded by residential and commercial areas.

On the date of testing, WWTP No. 10 was in the process of being upgraded to full secondary treatment, as part of the flow (approximately 49%) was being conveyed directly from the primary settling tanks to the ocean outfall. Full upgrade was completed in October 2002. This process train is referred to in this study as the "South Plant." The remaining 51% of the flow is treated in the "North Plant," undergoing full secondary treatment.

The WWTP influent is first treated in the headworks with bar screens, grinders, and aerated grit chambers. Prior to entering the primary tanks, the flow is dosed with an anionic polymer. After treatment in primary settling tanks, the flow is split between the secondary treatment process in activated sludge reactors and the flow that passes through traveling screens and then is pumped to the ocean outfall. The secondary treatment process consists of activated sludge reactors and secondary settling tanks. Chlorine is added to the liquid of both process trains prior to the effluent being sent to the outfall, which extends two miles into the ocean to a depth of 200′.

The solids processes for WWTP No. 10 are depicted in Figure 3-10. In the North Plant, WAS from secondary settling tanks is thickened in two DAFTs, which operate 24 hours a day. TWAS (Sample Location B) is produced at a rate of approximately 0.5 mgd, at a solids concentration of about 7%. Polymer is not used in the thickeners.

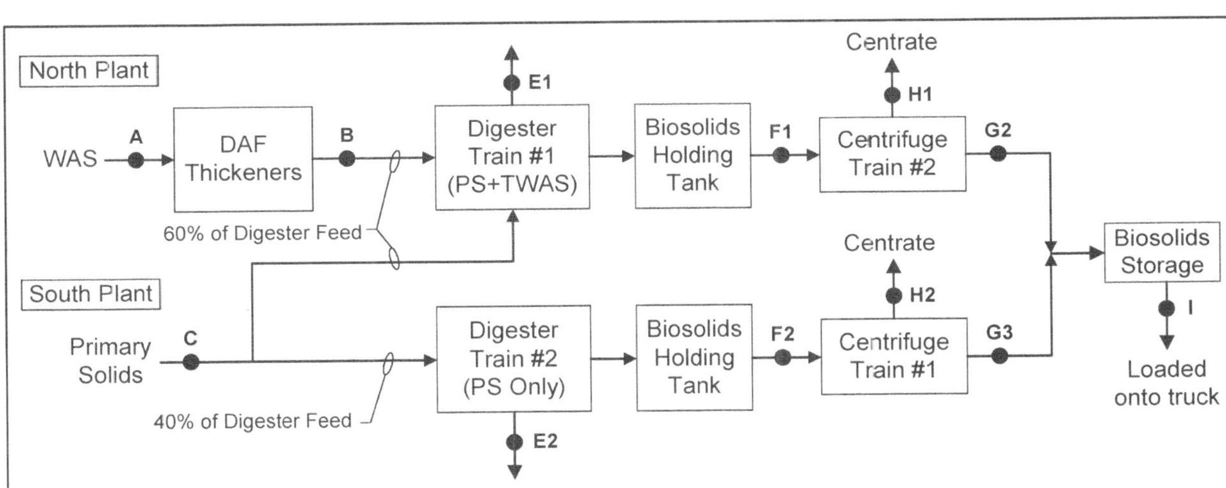

Figure 3-10. Schematic of WWTP No. 10.

All solids (TWAS) from the North Plant pass through the first solids train (Figure 3-10) and enter 11 mesophilic digesters, all of which have fixed covers. Those digesters also receive a portion of the PS from the South Plant. The total sludge feed to the North Plant digesters during the sampling was approximately 2.2 mgd (about 60% of the total digester feed). The second solids train at the South Plant contains 23 mesophilic digesters, all of which have fixed covers, and only treat PS. The total sludge feed to the South Plant digesters during the sampling was approximately 1.7 mgd, which is approximately 40% of the total digester feed. All digesters provide between 18–22 days of solids retention time. Iron salts are added to the primary sludge only prior to the split between the North Plant and South Plant digesters.

All digested sludge is dewatered at WWTP No. 10 using centrifuges. There are 10 low-solids centrifuges (approximately 1,400 rpm) available for dewatering at the South Plant, all of which were on line at the time of the testing. The South Plant centrifuges operate 24 hours a day and typically produce 25-30% solids cake. Polymer is added to the centrifuges to aid in dewatering.

At the North Plant, 25 centrifuges are available for dewatering, one of which is a high-solids centrifuge and the remainder of which are medium-solids centrifuges identical to those at the South Plant. The North Plant centrifuges operate 24 hours a day and use polymer as a dewatering aid. During the planning of the testing for WWTP No. 10, the project team decided to sample both the high-solids and medium-solids centrifuges at the North Plant; however, the high solids centrifuge was not on line during the date of testing. Therefore, there was no Sample Location G1. Samples G2 and G3 contained dewatered cake from medium-solids centrifuges at the North Plant and South Plant, respectively. The corresponding filtrate samples were H1 and H2.

Cake from both trains (Sample Locations G2 and G3) is then combined and conveyed via a conveyor belt into a silo, where it is stored for approximately one day. Sample Location I represents cake from the silo, stored for approximately one day. After storage, the cake is hauled away in trucks for land application or composting.

3.10.2 General Summary of Results

Table 3-29 presents the results of the field tests on samples collected in Sample Locations A through I (there is no Sample Location D) for WWTP No. 10.

Table 3-29. Field Testing Results from WWTP No. 10.

Sample Location	Composite or Grab	Temperature (°C)	pH	ORP (mV)
A WAS	Composite[1]	29	6.7	-102
B TWAS	Composite[1]	27	6.8	-109
C Liquid PS	Composite[1]	29	5.6	-50
F1 Train #1 Digested Sludge	Grab	33	7.4	-188
F2 Train #2 Digested Sludge	Grab	32	7.1	-196
G2 Train #1 Centrifuge Cake	Grab	32	--	--
G3 Train #2 Centrifuge Cake	Grab	32	--	--
H1 Train #1 Centrate	Grab	33	7.7	7
H2 Train #2 Centrate	Grab	34	7.6	83
I Stored Biosolids Cake	Grab	27	--	--

[1]Three composite samples, separated by one hour (results are average of three composites).

Table 3-30 presents the results of the field headspace analysis on samples collected during WWTP No. 10 testing. Duplicates were performed on several H_2S and NH_3 measurements.

Table 3-30. Field Headspace Testing Results from WWTP No. 10.

Sample Location	H2S Concentration (ppmv)	NH3 Concentration (ppmv)
INF Influent	5.15	0
A WAS	0.13	0
B TWAS	0.12	0.03
C Liquid PS	1.6	0
E1 Train #1 Digester Gas	25	1
E2 Train #2 Digester Gas	25	2.5
F1 Train #1 Digested Sludge	0.14	50
F2 Train #2 Digested Sludge	0.18	20
G2 Train #1 Centrifuge Cake	2.5	20
G3 Train #2 Centrifuge Cake	2.3	6.8
H1 Train #1 Centrate	1.0	23
H2 Train #2 Centrate	0.09	11
I Stored Biosolids Cake	4.9	26

Table 3-31 presents the results of the odor analysis on samples collected for WWTP No. 10. On August 22, 2002, one week after sampling, St. Croix Sensory measured for odor intensity in the headspace gas from the Train #2 digested sludge sample (Sample Location F2) and all cake samples (Sample Locations G2, G3, and I).

Table 3-31. Odor Evaluation Results from WWTP No. 10.

Sample Location	Detection Threshold (D/T)	Recognition Threshold (R/T)
F2 Train #2 Digested Sludge	1,300	730
G2 Train #1 Centrifuge Cake	12,000	8,100
G3 Train #2 Centrifuge Cake	8,700	5,700
I Stored Biosolids Cake	21,000	11,000

Note:
Samples were diluted 1:50 with nitrogen prior to shipment to St. Croix Sensory Laboratory; detection and recognition thresholds of the headspace gases in the sample bottles are 50 times the values reported in this table.

3.10.3 General Observations

WWTP No. 10 is one of the largest WWTPs in the study, and has several unique aspects. The following were noted by the project team:

- WWTP No. 10 contains one of the larger industrial components (approximately 20%) of its influent wastewater of all WWTPs in the study.
- WWTP No. 10 offered a unique experience for the project team, as the state of the WWTP during the time of sampling produced in effect two separate liquid and solids

process trains. Odor and odorous compound levels were comparable from one train to the next.

- During the testing period, WWTP No. 10 contained digesters that were used for treating PS only (at the South Plant) and digesters that were used for treating combined primary and TWAS (at the North Plant), which allows for a direct comparison of odorous constituents for each train.
- WWTP No. 10 is one of two WWTPs in the study that uses a high-solids centrifuge (WWTP No. 2 is the other); unfortunately, that centrifuge was not on line during the sampling.
- WWTP No. 10 contained some of the highest influent field-tested headspace concentrations of H_2S in the study, averaging 5.2 ppmv.
- Thickening of the WAS had little effect on the headspace H_2S and NH_3 concentrations measured in the field. This is similar to what was observed in WWTP No. 9, but different from WWTP No. 3, where TWAS produced a lower H_2S headspace concentration, and WWTP No. 8, where thickening WAS produced a slightly higher H_2S headspace concentration.
- The North Plant digested sludge (Sample Location F1, effluent of the digesters that treated a combination of TWAS and PS) contained the highest NH_3 measurements of all WWTP No. 10 samples. The average NH_3 field headspace concentration of 50 ppmv was more than twice that of Sample Location F2, where sludge from the South Plant digesters originated only as PS.
- A similar effect was noted in the Train #1 cake odor measurements (Sample Location G2), which were higher than the Train #2 odor measurements (Sample Location G3).
- Field headspace H_2S and NH_3 concentrations of the dewatered cake increased following combination of the two trains and after one day of storage (between Sample Locations G2/G3 and I).
- Similar to other WWTPs, dewatering of digested biosolids produced much higher odor readings in post-dewatering samples compared to pre-dewatering samples (order of magnitude increase between samples F2 and G2).
- Similar to the field headspace analysis results, cake from Sample Location I (following combination of the two trains and one day of storage) contained higher odor than cake from Sample Locations G2 or G3.

3.11 WWTP No. 11

3.11.1 WWTP Description

WWTP No. 11 treats wastewater generated in a large city. The facility influent is approximately 93% domestic wastewater, with the remaining 7% a combination of wastewater from food industries discharging into the collection system.

The WWTP has a nominal capacity of approximately 216 mgd, and on the day of testing, August 23, 2002, the influent flow was 200 mgd. Ferrous chloride is introduced to the wastewater upstream of the WWTP for phosphorus removal. Two influent pump stations deliver the WWTP influent, which passes through two separate preliminary/primary treatment trains.

The first train has six aerated grit chambers, which remove collected grit by a clam bucket system. The grit is landfilled. The wastewater is distributed from the grit chambers to nine primary settling tanks. Six of the tanks are covered and equipped with chain and flight mechanisms. The remaining three are open and contain sludge collection bridges. The second preliminary/primary treatment train has four aerated grit chambers, followed by three primary settling tanks equipped with sludge collection bridges.

Primary effluent is collected in a common conduit and then distributed into 11 covered step feed-activated sludge basins with coarse bubble aeration. The mixed liquor in each individual aeration tank is sent to an associated secondary clarifier tank. Secondary clarifier effluent is disinfected and then discharged into Lake Ontario via an outfall pipe and diffuser system.

Figure 3-11 depicts the solids flow schematic for WWTP No. 11, along with each sample location. WAS is thickened in 10 DAF tanks to 3-5% total solids content. Mixed primary sludge and TWAS are anaerobically digested in 16 fixed-roof mesophilic digesters. Digester contents are re-circulated by pumps and digested biosolids are exposed to a second stage of digestion in four secondary tanks.

Biosolids are dewatered in 12 high-solids centrifuges. Dewatered biosolids with approximately 30% solids content are either transferred to a biosolids drying facility or to a cake storage area for land application. Biosolids are transferred between the centrifuges and the drying facility or storage area via high pressure pumps. The biosolids drying and pelletizing facility contains two trains with a biosolids storage tank, a multi-tray dryer, and a truck loading silo.

Approximately 2.2 mgd of digested sludge flows from the primary digesters past Sample Location F1 (Figure 3-11) to the secondary digesters, which provide four additional days of VS destruction. The Sample Location F2 represents the end of the digestion process, following the day tanks. Polymer is added in front of the centrifuges to aid in dewatering. Sample Locations G and I represent post-centrifuge samples, with Sample Location G being prior to high pressure pumps and Sample Location I representing cake quality after cake pumping.

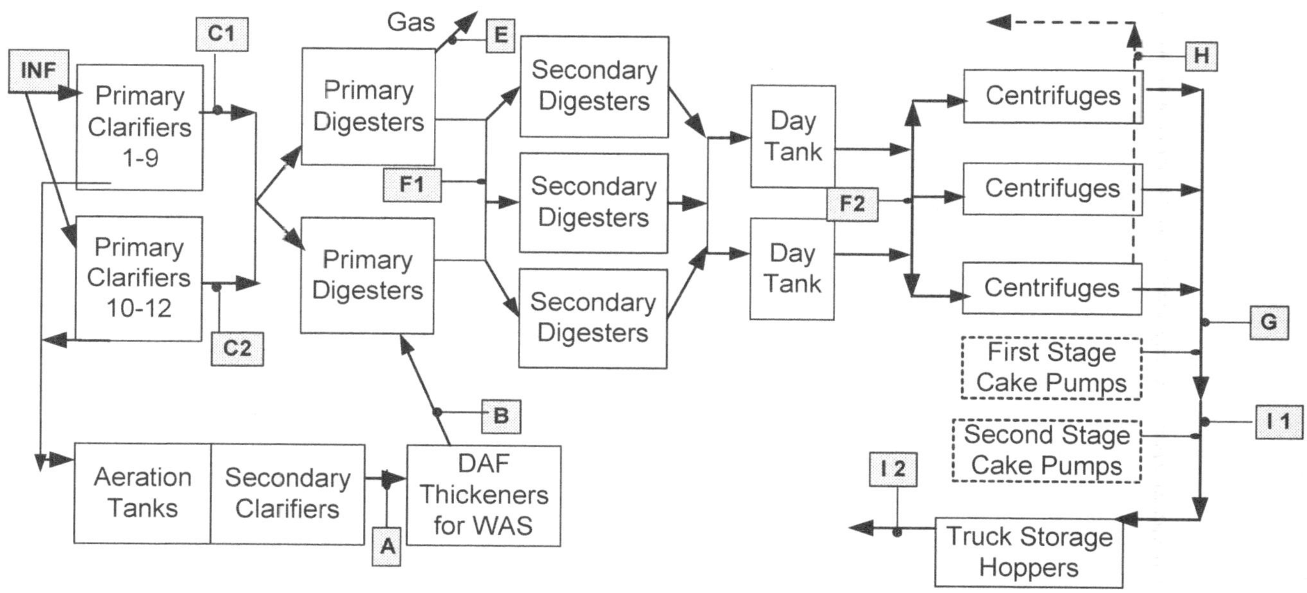

Figure 3-11. Schematic of WWTP No. 11.

3.11.2 General Summary of Results

Table 3-32 presents the results of the field tests on samples collected in Sample Locations A through I2 (there is no Sample Location D) for WWTP No. 11.

Table 3-32. Field Testing Results from WWTP No. 11.

Sample Location	Composite or Grab	Temperature (°C)	pH	ORP (mV)
INF Influent	Composite	11	7.8	147
A WAS	Composite[1]	24	6.2	-104
B TWAS	Composite[1]	24	6.5	-138
C1 Liquid PS 1	Composite[1]	24	5.9	-39
C2 Liquid PS 2	Composite[1]	23	5.7	-126
F1 Secondary Digested Sludge	Grab	38	7.0	-175
F2 Primary Digested Sludge	Grab	36	7.1	-175
G Centrifuged Biosolids	Grab	39	8.0	23
H Centrate	Grab	36	7.5	85
I1 Cake Post-Conveyance	Grab	42	7.5	-148
I2 Cake Post-Storage	Grab	26	7.5	-155

[1]Three composite samples, separated by one hour (results are average of three composites).

Data from the headspace field testing at WWTP No. 11 are presented in Table 3-33. Duplicates were performed on several H_2S and NH_3 measurements. The averages of all measurements are reported in Table 3-33.

Table 3-33. Field Headspace Testing Results from WWTP No. 11.

Sample Location	H_2S Concentration (ppmv)	NH_3 Concentration (ppmv)
INF Influent	0.04	<0.2
A WAS	0.25	<0.2
B TWAS	1.1	<0.2
C1 Liquid PS 1	1.1	<0.2
C2 Liquid PS 2	10.6	<0.2
E Digester Gas	1.5	<0.2
F1 Secondary Digested Sludge	0.2	4
F2 Primary Digested Sludge	0.08	14
G Centrifuged Biosolids	0.01	17
H Centrate	0.01	40
I1 Cake Post-Conveyance	0.97	3
I2 Cake Post-Storage	0.11	55

Table 3-34 presents the results of the odor analysis on samples collected for WWTP No. 11. On August 30, 2002, one week after sampling, St. Croix Sensory measured for odor intensity in headspace gas from all sampling points downstream of digestion, with the exception of Sample Location H – dewatering centrate samples. Detection and recognition thresholds were measured as part of the olfactometry analysis.

Table 3-34. Odor Evaluation Results from WWTP No. 11.

Sample Location	Detection Threshold (D/T)	Recognition Threshold (R/T)
F2 Primary Digested Sludge	100	65
G Centrifuged Biosolids	15,000	8,700
I1 Cake Post-Conveyance	13,000	8,700
I2 Cake Post-Storage	1,300	730
Note: Samples were diluted 1:50 with nitrogen prior to shipment to St. Croix Sensory Laboratory; detection and recognition thresholds of the headspace gases in the sample bottles are 50 times the values reported in this table.		

3.11.3 General Observations

The following observations were noted by the project team:

- With the exception of Sample Location C2 (Liquid PS 2), all samples contained relatively low concentrations of H_2S.
- All sample points prior to digestion contained NH_3 levels that were below the detection limit.
- The digester gas (Sample Location E) contained low levels of H_2S and below detection limit NH_3.
- Similar to other WWTPs, odor measurements increased upon dewatering of the sludge (Table 3-34, Sample Location G).

- Conveyance of the dewatered biosolids appeared to have minimal effect on total odor.
- Following storage, the dewatered cake headspace measurements showed a decrease in H_2S and a sharp increase in NH_3.
- Following storage, the total odor measurements decreased in the biosolids cake by an order of magnitude (Table 3-34). Prior studies at WWTP No. 11 had found an increase in odors from biosolids cake after storage, along with relatively high concentrations of nitrogen-based compounds indole and skatole.

CHAPTER 4.0

RESEARCH FINDINGS

Chapter 4.0 summarizes the findings of the Phase II research and presents conclusions and recommendations. The chapter begins with a discussion of the chemical compounds that create odors in biosolids (Subsection 4.1) and then discusses the sampling and analytical methods that were used in the study for testing the odorous compounds emitted from the biosolids (Subsection 4.2). Following the discussion of biosolids odors and how they are measured, the chapter covers the constituents in wastewater and biosolids that are precursors to odors (Subsection 4.3).

Subsection 4.4 discusses the potential impacts of wastewater and biosolids processes on odors in general order of the flow of wastewater and biosolids through a typical wastewater treatment facility, as follows:

- Subsection 4.4.1 – Impacts of processes upstream of anaerobic digestion
- Subsection 4.4.2 – Impacts of anaerobic digestion
- Subsection 4.4.3 – Impacts of biosolids dewatering and conveyance processes
- Subsection 4.4.4 – Impacts of biosolids cake storage

Each of the subsections follows the same general format: 1) a hypothesis; 2) a summary of results pertaining to the hypothesis; 3) a discussion of the results; and 4) conclusions and recommendations with respect to the original hypothesis. During the preparation of this chapter many potential relationships between treatment parameters and observed sensory and chemical odor measurements were considered, but only a fraction of these are presented in this part of the report. The ones omitted did not show any correlation.

Throughout Chapter 4.0 the numerical identifiers (1 through 11) associated with various data points refer to their respective WWTP sources. Alpha character identifiers (A through I) refer to their respective Sample Locations. Generalized Sample Location identifiers are shown in Figure 2-1 in Chapter 2.0, and specific Sample Location identifiers for each WWTP tested are shown on each WWTP schematic presented in Chapter 3.0.

4.1 Odorous Compounds in Biosolids

Volatile sulfur compounds (VSCs) are known to contribute significantly to odor problems of digested biosolids cake produced by centrifuges (Higgins et al., 2003; Murthy et al., 2002; Novak et al., 2002). Trimethylamine (TMA) is a nitrogen-based compound often associated with a fishy odor in limed sludges. Indole and skatole are odorous aromatic amines (also nitrogen-based) that were first found in mammalian feces and could cause a fecal odor scent. Fatty acids are common in biosolids and produce a rancid smell.

Odor is defined in this study as a human perception that can be quantified by olfactometry in odor units of dilutions-to-threshold (D/T) and in quantitative terms of Detection Threshold (DT), the number of Odor Units (D/T) at which an odor is detected, or Recognition Threshold (RT), the number of Odor Units (D/T) at which an odor is recognizable by descriptive terms. Odor can also be qualified in descriptive terms, such as pungent, rancid, fecal, and rotten.

The odors of selected biosolids samples have been both quantified and qualified at each of the 11 test WWTPs. The Phase II research began with the hypothesis that the odor of biosolids is caused by volatile chemicals that can be measured in the headspace of biosolids in bottles. The project team also hypothesized a correlation between the concentration of odorous compounds and quality and quantity of odors. The hypothesized odor-causing compounds analyzed by chemical odor methods (GC/MS) were as follows:

- The sulfur compounds H_2S, methanethiol or methyl mercaptan (MT), dimethylsulfide (DMS), dimethyldisulfide (DMDS), dimethyltrisulfide (DMTS), carbonylsulfide (COS), carbondisulfide (CS_2). As a group, they are referred to as VSCs.
- The nitrogen compound TMA.
- The nitrogen compounds indole and skatole, which are aromatic amines.
- Fatty acids, which are odorous but difficult to measure in headspace. These were measured by direct liquid analysis at Bucknell University to compare these results with olfactometry measurements.

Odors themselves were measured by olfactometry, using human test panels that worked with the same headspace gas samples used for chemical analysis. However, olfactometry was undertaken only with headspace gas taken at Day 6 of incubation, which was indicated by prior research as the time period required to generate maximum odor levels from biosolids samples.

4.1.1 Results and Discussion

Olfactometry data for the 11 WWTPs that took part in this study are listed in Table 4-1.

Table 4-1. Results of Olfactometric and Chemical Measurements Performed on Sample Bottle Headspace Gas Samples.

WWTP No.	Sample Location	Detection Threshold[1] (DT)	Recognition Threshold[1] (RT)	Peak Total Sulfur mg S/m^3	Peak Total Nitrogen mg N/m^3
1	F2 Digested Biosolids	360	230	4.0	ND[5]
	G Fresh Low-Solids Centrifuge Biosolids Cake	17,000	11,000	2020[2]	ND
	I2 Low-Solids Centrifuge Cake after about 7-10 days storage	18,000	14,000	1774[3]	ND
2	F2 Digested Solids After Holding Tank	390	230	4.0	ND
	G Low-Solids Centrifuge Biosolids Cake	6,100	4,300	352	ND
2	I High-Solids Centrifuge Biosolids Cake	21,000	13,000	787	ND
3	F Digested Biosolids	460	270	4.8	ND
	G Dewatered Biosolids	9,600	7,300	416	ND
	I Stored Biosolids Cake	4,800	4,200	173	ND
4	F2 Digested Sludge	230	120	5.0	ND
	G Lagoon Top Biosolids Sample	3,700	1,600	60	ND
	I Lagoon Top Biosolids Sample	3,500	2,000	27	ND
5	F Digested Biosolids	270	140	2.7	ND
	G High-Solids Centrifuge Biosolids Cake	6,100	3,500	494	1.1
	I High-Solids Centrifuge Cake after about 2 days storage	7,400	4,300	394	1.7
6	F1 Digested Biosolids (DS)	95	70	8.0	ND
	F2 DS After Holding Tank	120	75	1.0	ND
	G Fresh Biosolids Cake	5,100	3,100	139	1.0
	I Stored Biosolids Cake	2,900	1,700	131	1.0
7	F Digested Biosolids	1,300	830	19.3	ND
	G Centrifuge Biosolids Cake	19,000	14,000	2408[4]	4.34
	I Drying Bed Biosolids Cake	1,900	1,400	67	0.56
8	F2 DS Post-Screening	120	65	2.8	1.0
	G Centrifuge Biosolids Cake	9,100	5,000	621	2.23
	I1 Cake Post-Conveyance	2,500	1,400	578	3.33
	I2 Cake Post-Storage	8,900	6,100	304	2.85
9	F1 Digested Sludge	2,500	1,400	14.3	ND
	F2 Digested Sludge Post-Storage	95	70	1.8	ND
	G Plate & Frame Filter Press Cake	1,700	1,100	19	0.97
	I Cake Stored for Two Days	2,200	1,300	130	0.85
10	F2 Train #2 Digested Sludge	1,300	730	4.3	ND
	G2 Train #1 Centrifuge Cake	12,000	8,100	874	0.91
	G3 Train #2 Centrifuge Cake	8,700	5,700	632	1.13
	I Stored Biosolids Cake	21,000	11,000	1160	0.87
11	F2 Primary Digested Sludge	100	65	0.7	ND
	G Centrifuged Biosolids	15,000	8,700	819	0.54
	I1 Cake Post-Conveyance	13,000	8,700	983	0.72
	I2 Cake Post-Storage	1,300	730	19	2.14

Notes: Olfactometry samples were collected on Day 6 of storage. Values for DT and RT are in Odor Units (D/T). Peak total sulfur and nitrogen concentrations were not necessarily measured on samples collected on the 6th day of storage.

[1] These measurements were performed on gas samples obtained from headspace bottles on Day 6 of storage and collected in Tedlar® bags, without any headspace losses during the 6-day storage period. Day 6 was chosen, since the days-to-peak values were not known at the time of sampling. The values represent a 1:50 dilution in samples.

[2] This value represents a 93% H_2S contribution to total sulfur concentration.

[3] This value represents an 89% H_2S contribution to total sulfur concentration.

[4] This value represents a 58% H_2S contribution to total sulfur concentration.

[5] "ND" means "not detected," results were below the analytical detection limit.

Table 4-2 lists the olfactometry and headspace chemical odor measurement results ranked based on the odor detection thresholds measured in the dewatered cake (G-cake) odor values. It shows that WWTP No. 9 cake was least odorous and WWTP No. 2H (high-solids centrifuge train) was the most odorous based on the sensory odor measurements. Once the "Detection Threshold" and "Peak Total Sulfur" columns are compared, it is evident that the latter parameter follows the same order as the former. In other words, a strong relationship between the headspace odor and peak total sulfur is implicit from this data. The nitrogenous odor compounds measured in the headspace were not high enough to indicate a correlation.

Table 4-2. Results of Olfactometric and Chemical Measurements Performed on G-Cake Bottle Headspace Gas Samples Ranked According to Detection Threshold Values.

WWTP No.	Sample Location	Detection Threshold[1] (DT)	Recognition Threshold[1] (RT)	Peak Total Sulfur mg S/m^3	Peak Total Nitrogen mg N/m^3
2H	I High-Solids Centrifuge Biosolids Cake	21,000	13,000	787	ND[4]
7	G Centrifuge Biosolids Cake	19,000	14,000	2408[3]	4.34
1	G Fresh Low-Solids Centrifuge Biosolids Cake	17,000	11,000	2020[2]	ND
11	G Centrifuged Biosolids	15,000	8,700	819	0.54
10	G2 Train #1 Centrifuge Cake	12,000	8,100	874	0.91
3	G Dewatered Biosolids	9,600	7,300	416	ND
8	G Centrifuge Biosolids Cake	9,100	5,000	621	2.23
10	G3 Train #2 Centrifuge Cake	8,700	5,700	632	1.13
2	G Low-Solids Centrifuge Biosolids Cake	6,100	4,300	352	ND
5	G High-Solids Centrifuge Biosolids Cake	6,100	3,500	494	1.1
6	G Fresh Biosolids Cake	5,100	3,100	139	1
4	G Lagoon Top Biosolids Sample	3,700	1,600	60	ND
9	G Plate & Frame Filter Press Cake	1,700	1,100	19	0.97

Notes: Olfactometry samples were collected on Day 6 of storage. Values for DT and RT are in Odor Units (D/T). Peak total sulfur and nitrogen concentrations were not necessarily measured on samples collected on the 6th day of storage.

[1] These measurements were performed on gas samples obtained from headspace bottles on Day 6 of storage and collected in Tedlar® bags, without any headspace losses during the 6-day storage period. Day 6 was chosen since the days-to-peak values were not known at the time of sampling. The values represent a 1:50 dilution in samples.

[2] This value represents a 93% H_2S contribution to total sulfur concentration.

[3] This value represents a 58% H_2S contribution to total sulfur concentration.

[4] "ND" stands for "not detected," results were below analytical detection limit.

Figures 4-1 and 4-2 show that the maximum concentration for volatile nitrogen in analyzed samples was generally on the order of 1,000 times lower than the maximum concentration for volatile sulfur in the same samples. These figures also demonstrate that most of the odor samples analyzed from the test WWTPs show a positive correlation between olfactometry measurements in terms of odor DT and volatile sulfur in concentration units of milligrams of sulfur per cubic meter (mg S/m^3).

Figure 4-2 indicates no discernible relationship between odor DT and volatile nitrogen (milligrams of nitrogen per cubic meter [mgN/m^3], measured as TMA, indole, and skatole). This has two implications: nitrogen compounds are secondary odor producers compared to sulfur compounds, and the GC/MS method needs to be improved to better capture the N-bearing odorous compounds.

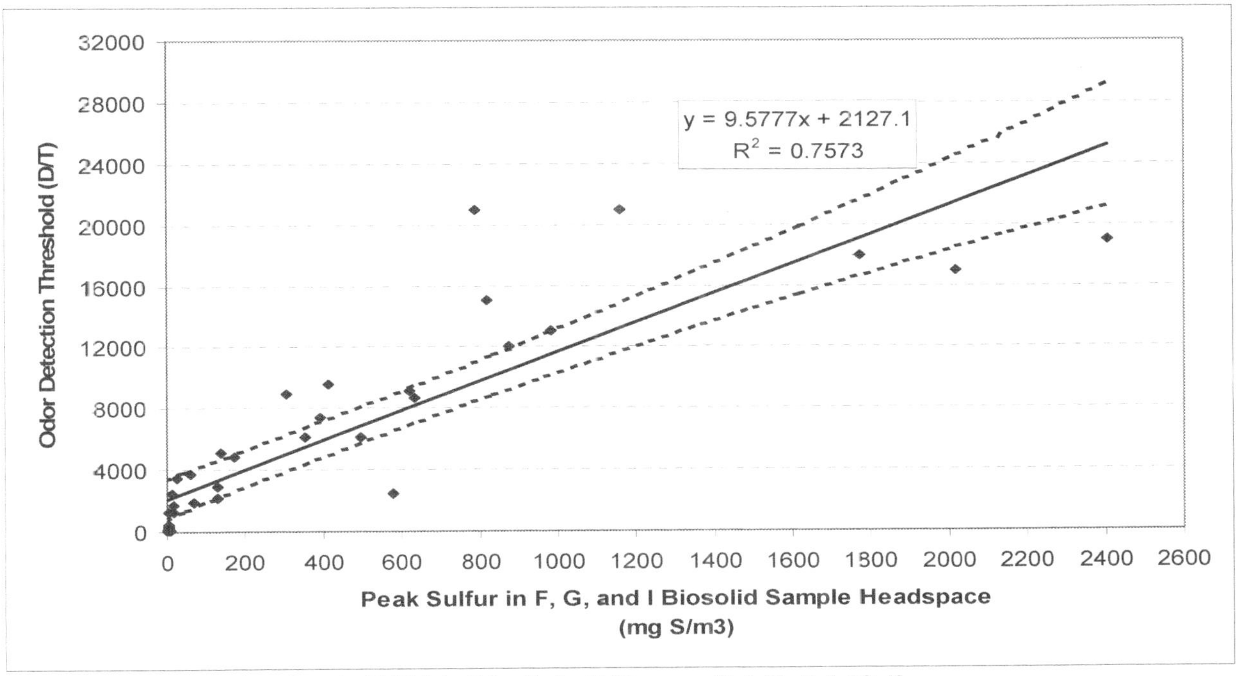

Figure 4-1. Odor Detection Threshold (DT) in Odor Units (D/T) versus Volatile Total Sulfur.
(Dashed Lines Represent 95% Confidence Interval).

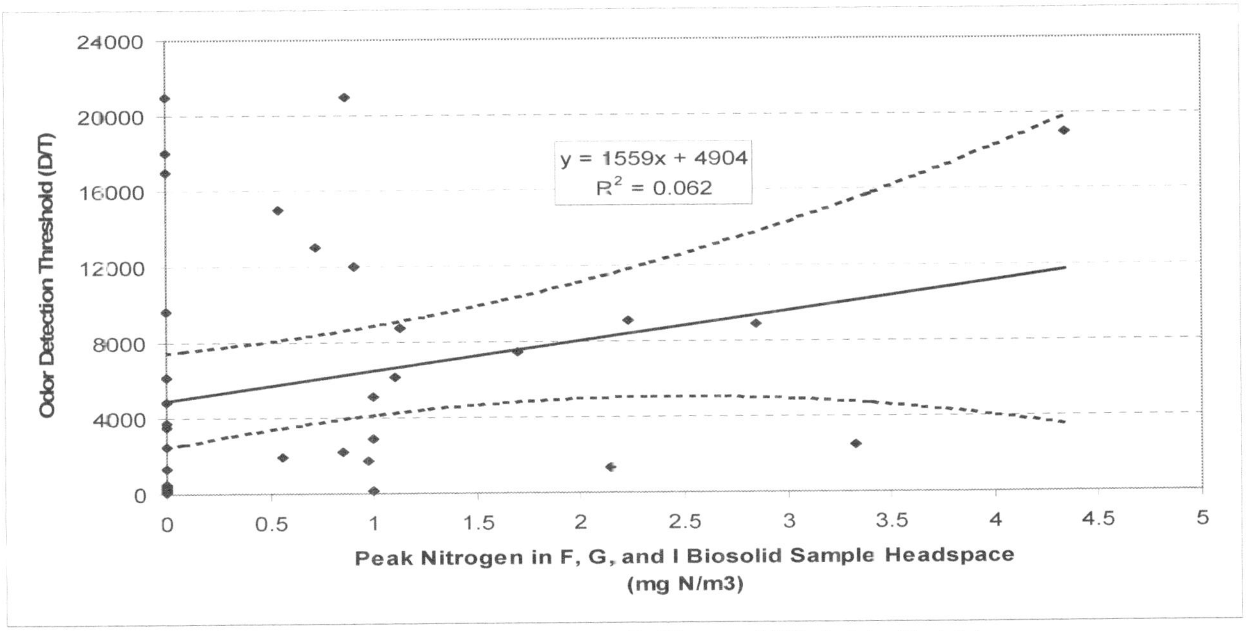

Figure 4-2. Odor Detection Threshold (DT) versus Peak Volatile Nitrogen (TMA, Indole, Skatole).
(Dashed Lines Represent 95% Confidence Interval).

Figure 4-3 indicates a strong relationship between olfactometry measurements expressed as odor DT and MT with a correlation coefficient (R^2) of 0.72. WWTP No. 1 did not fit this correlation, likely because it contained significantly higher H_2S than the other WWTPs, as indicated in Table 4-1. High H_2S resulted in a high odor level, but, since most of the sulfur was in the form of H_2S, MT values were low at this WWTP. WWTP No. 4 was not included in the correlation because its cake was not mechanically dewatered, but was stored in a lagoon.

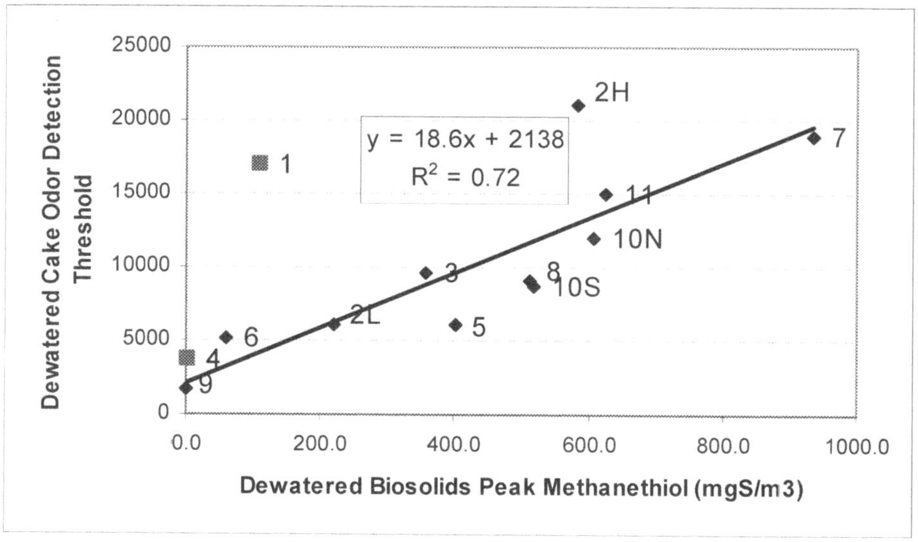

Figure 4-3. Peak MT versus Odor Detection Threshold (DT).

Figure 4-4 illustrates the absence of a correlation between olfactometry measurements on post-digestion samples and fatty acids concentrations in the liquid phase of digester effluent. VFAs were expressed as mg/bottle (mg of VFA in mg of wet sample in the bottle), which reflects the mass of fatty acids per wet mass of sample. Fatty acids in the gas phase were also estimated with the headspace method and GC/MS but were not calculated because of a lack of confidence in current analytical methods to accurately quantify fatty acids. Nevertheless, based on the analyses, it can be confidently estimated that the total concentration of fatty acids was not higher than 50 ppm in any of the biosolids samples.

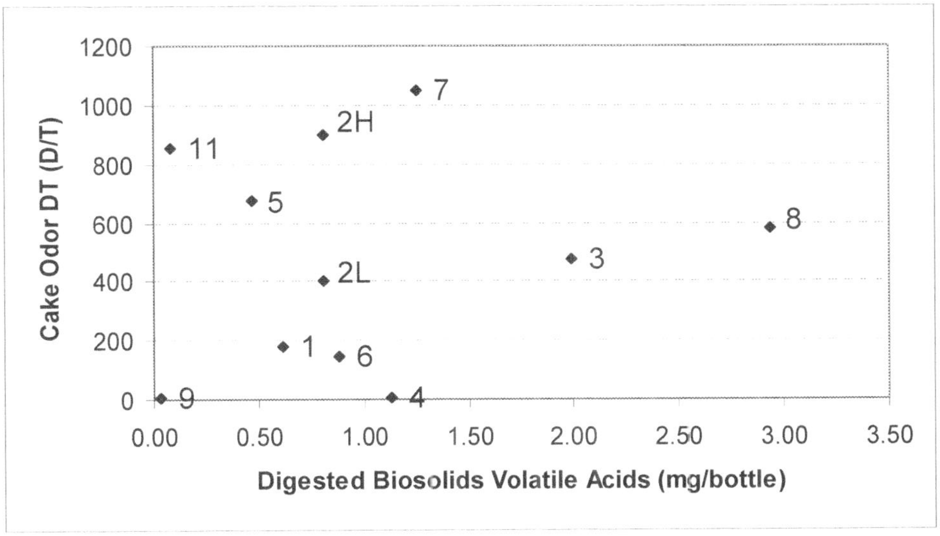

Figure 4-4. Odor Detection Threshold (DT) on Post-Digestion Biosolids Samples versus VFAs in Liquid Digested Biosolids (Agency Data).

A multiple regression analysis showed that the volatile sulfur compounds (MT, DMS, DMDS, and DMTS) were very well correlated with the odor measurements on Day 6 as measured by olfactometry (DT). The correlation coefficient for this relationship is 0.90. This data support findings that show VSCs are responsible for much of the odor, especially in the first week of cake storage. The equation for this relationship is as follows:

$$DT = 3048 + 3.2(MT) - 29.7(DMS) + 105(DMDS) + 1060(DMTS)$$

where:

DT = detection threshold on Day 6;
MT = peak methanethiol concentration (mg S/m^3);
DMS = peak dimethyl sulfide concentration (mg S/m^3);
DMDS = peak dimethyl disulfide concentration (mg S/m^3);
DMTS = peak dimethyl trisulfide concentration (mg S/m^3).

Interestingly, the DMS concentration appears to reduce odors measured by DT as shown by the negative term on the coefficient. This contradicts some previous research that showed DMS as an important odor source, although the complex interactions of the different species in the model might have contributed to these results.

The hedonic odor descriptors listed in Table 4-3 also indicate that the offensive odors of VSCs are likely the dominant odor from anaerobically digested biosolids. A fishy odor was rarely reported in the anaerobically digested biosolids from any of the WWTPs, which tends to substantiate the low to non-existent concentrations of TMA found by parallel chemical analysis. On the other hand, "offensive" characteristics were reported in the biosolids cakes of every WWTP tested, which foretells why odors are the primary issue that affects the public acceptance of biosolids land application at present.

Odor descriptors are subjective measures that are based on the perception of a trained odor panel and cannot be used to deduce the chemical source of the odor. Different compounds have different limits of detection by the human nose, which also varies based on a person's RT for a specific scent. For example, the odor detection limit for DMDS is reported as 0.1

micrograms per cubic meter ($\mu g/m^3$), for DMS as 2.5 $\mu g/m^3$, and for NH_3 as 26 $\mu g/m^3$ (Rosenfeld et al., 2001). When these odorous compounds are present as a mixture, their odor characteristics, as well as detection and recognition thresholds, are affected. For this reason the olfactory measurements in terms of DT, RT, and qualitative descriptors are all valuable when investigated in conjunction with the GC/MS results.

Table 4-3. Percentage of Panelists Reporting Presence of Each Odor Descriptor for Tested Dewatered Cake Samples (G-Cakes) Obtained from Each Test WWTP.

Descriptor	No. 1	No. 2	No. 3	No. 4	No. 5	No. 6	No. 7	No. 8	No. 9	No. 10	No. 11
Chemical	40			40		60	40		79	79	
Car Exhaust				40		60				40	
Decay		40	40		40		60	40			
Earthy	60	60			60		60	60			
Fecal	60	40				60	40				
Fishy			40								
Garbage		40	60		79		60				40
Gasoline				40		40			40		
Manure	40										
Musty							40				
Offensive	100	100	100	79	100	100	100	100	100	100	79
Putrid	60	60	79	60	60	40	60				60
Rancid		40	79		60		79				
Raw meat							40				
Rotten eggs						40		40	40	40	
Septic	40					40				40	40
Sewer	40		40	79	40	60					40
Sour	40	40	60		40		40				60
Stale								40			
Sulfur						40					
Swampy	60				40						
Vegetable			40								40

Multiple regression analysis of cake odor and nitrogenous compounds (mg N/m3) or Volatile Acids (mg/L) did not reveal any meaningful and statistically significant correlation. However, it must be remembered that the olfactometry analyses were performed on dewatered cake, and the potential impact of mechanical dewatering on biosolids physical and chemical properties might have contributed to the absence of correlation. Also, nitrogen-based odors may either be masked by the sulfur odors during the first week of storage or be more prominent after a longer storage period.

4.1.2 Conclusions

The following conclusions are based on the results described above:

- The comparison of olfactometry with chemical analyses supports the hypothesis that odors from biosolids (a subjective human perception) can be quantified by olfactometry and correlated with the concentrations of odorous chemical compounds.
- Chemical and olfactory odor analyses on Day 6 of incubation reveal that odors from biosolids correlate well with volatile sulfur concentration (mg/m^3) in the sample headspace.

- Dewatered biosolids cake is still odorous after the volatile sulfur odor has disappeared. In hedonic terms, this residual odor is different from the sulfur odor and may be caused by other odorous chemicals, such as fatty acids, phenol, p-cresole, indole, or skatole. These compounds are difficult to analyze by chemical or conventional olfactometry methods, which could result in under estimation of this residual odor.

4.1.3 Recommendations

Based on the conclusions in Subsection 4.1.2, the project team made the following recommendations:

- Chemically analyze for VSCs, including H_2S, MT, DMS, and DMDS, as the dominant odorous compounds of biosolids within one to two weeks of dewatering mesophilically digested biosolids. Analyze for nitrogen based compounds on samples stored longer than two weeks.
- Since the one thermophilicly digested sludge produced a different odor pattern than mesophilicly digested sludges (the odor was produced more slowly), additional studies should be conducted using thermophilicly digested sludges.
- Involve more sophisticated chemical and olfactometry methods in future research to understand more fully residual biosolids odors that are not VSCs (fatty acids, phenol, p-cresole, indole, skatole). Unlike VSC measurements, which produced an unprecedented amount of data for this study, the existing methods and data on the residual odors are insufficient to make a judgment on their constituents.

4.2 Headspace Method of Odorous Compound Sampling

4.2.1 Hypothesis

Odors from biosolids depend to some degree on operating conditions within the WWTP and biosolids handling processes, as well as environmental influences after biosolids are produced at the WWTP. One of the objectives of this study was to gain a better understanding of how different process parameters and environmental conditions affect odors from biosolids.

The specific conditions of operation at a WWTP affect the odors of biosolids produced at the facility, but the odors are dynamic in nature. Recently dewatered biosolids that have not been limed are more biologically active, and odorous compounds released from the biosolids can rise and decline within days, depending on how they are handled and stored after dewatering. Although the emission of odors from biosolids can be measured by chemical test tubes, sensors, or GC, these measurements are of limited use in finding the odor's cause.

In order to better understand the odors produced from biosolids, the project team performed bench-scale laboratory simulation tests under standardized conditions of biosolids storage and odor analysis. The project team selected the anaerobic headspace sampling method, which is capable of eliminating the changing conditions of emission flux, atmospheric dilution, and oxygen. These factors tend to make odors change rapidly, making it difficult to trace odors back to their sources in wastewater and process parameters.

The headspace sampling method also has other advantages over air sampling, flux chamber, and purge-and-trap sampling methods. Air sampling uses sampling bags (Tedlar®) or adsorption tubes (Tenax) to measure the odor concentration at the WWTP directly. Combined with chemical analysis, this method measures the concentration of odor compounds in air at a given time, but it is not useful for measuring the changes in biosolids odors over time. Flux chambers can be used for collection of odorous compound emission samples under a more realistic, dynamic environment, but the released compounds are highly diluted in the flux gas, the SPME enrichment slows the analysis to up to an hour per sample, and calibration of the method is difficult.

The purge-and-trap sampling method, which was provided by the laboratory used to sample digester gas in this study, purges the sulfur content completely out of the biosolids and into a sample bag for analysis. Purging is a slow process that requires up to 40 minutes per sample and results in dilution of the odorous compounds. It also removes the gases from contact with the biosolids, thereby reducing the opportunity for the microbes in the biosolids to transform the odorous compounds over time.

The headspace method, on the other hand, overcomes the drawbacks of the other methods by using a biologically and chemically inert bottle for the biosolids storage, and analyzing the headspace gas in the bottle for odor and chemical constituents. The bottles serve as containers for sample shipment, preserve samples in an anaerobic state (as in a cake pile or biosolids storage hopper), act as bench-scale cake incubators to simulate full-scale biosolids storage, and provide containment of headspace gases that characterize the odor of the biosolids sample. GC, test tubes, and gas sensors can all be adapted to headspace bottles to facilitate headspace analysis.

Headspace odor samples can be used for chemical odor analysis (using GC/MS, test tubes, and sensors) and olfactory odor analysis (using human test panels and an olfactometer). The headspace method of sampling and bottle storage provides fast and reliable odor analyses and, as a result, monitoring of the odor production-consumption cycles for all the samples.

4.2.2 Results

The headspace sample bottles were shown to be inert when GC analysis of limed (inactive) biosolids stored for six weeks in PET bottles resulted in a constant concentration of VSCs (Figure 4-5). These results also confirmed that the PET bottles used as part of the biosolids storage protocol did not cause any loss of headspace gases.

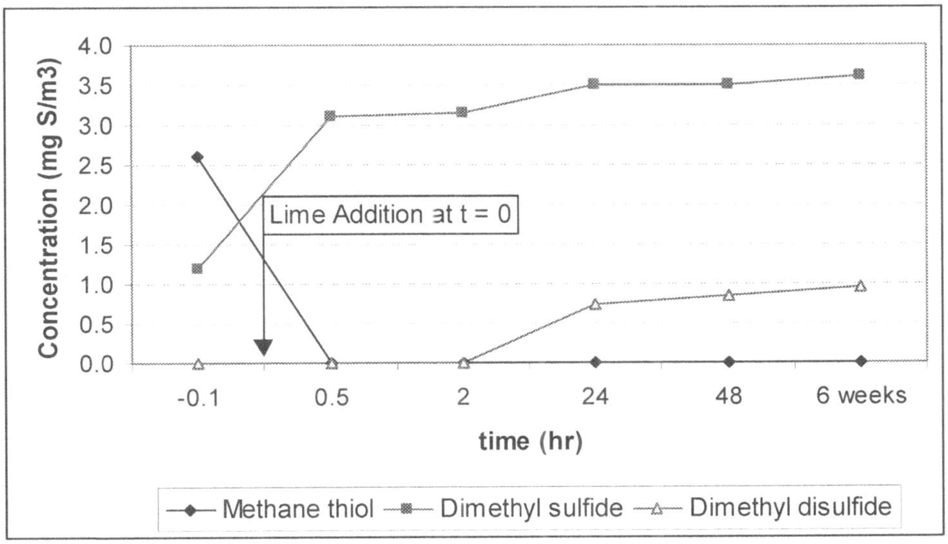

Figure 4-5. Concentration of Three VSC Compounds versus Storage Time of Inactive Biosolids Samples (Novak et al., 2002).

The validity of the method to simulate storage was investigated by measuring oxygen, CO_2, and CH_4 in the headspace of the bottles containing biosolids. The results showed that oxygen depleted rapidly on Day 1 of incubation. It appears that the odor-producing storage conditions in the headspace bottles are similar to anaerobic conditions in biosolids cake piles or storage vessels. Biosolids stored in piles, storage hoppers, trucks, or landfilled are anaerobic and black in color within the interior, but aerobic in the gray or brown surface layers which have minimal odor emissions due to loss of moisture. Figure 4-6 is a schematic drawing of a biosolids storage pile, with flux chamber sampling (left) compared to bottle headspace sampling (right).

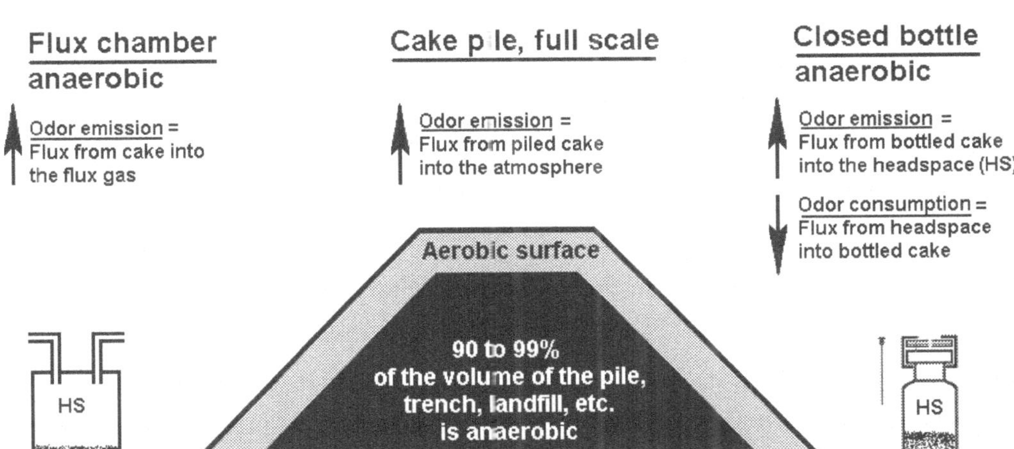

Figure 4-6. Anaerobic Versus Aerobic - Headspace Bottle versus Flux Chamber (Source: Glindemann, D. et al. 2004).

While it is recognized that odor emissions can be amplified by mixing and disturbing the stored piles and that aeration during mixing may even lead to different types of odors being generated, the researchers decided to focus on the undisturbed anaerobic conditions prevailing in most storage piles because it was believed to be a more practical and reproducible method for comparing the odor potential at different sampling locations at different WWTPs.

A standard incubation temperature of 22°C in the headspace bottles was used for this study because it is representative of average, ambient temperatures and is easy to maintain in the laboratory environment. Over the long term, the temperature of a biosolids cake generally rises above ambient levels due to the RBA in the cake. Although the temperature in biosolids storage piles or vessels can easily exceed 30°C in hot weather, which increases biological activity and gas production of biosolids, the aim of this study was to compare odorous compound emissions from different WWTPs when stored under identical conditions. Maintaining all the samples at the same temperature allowed for this type of comparison.

Figure 4-7 shows a stacked bar chart of the headspace concentration of four different VSCs measured on the indicated days of incubation, each of which was measured in an individual bottle dedicated for a specific day of storage for each sample for one WWTP. The top of each bar represents the total VSC concentration. The GC/MS measurements were performed in duplicates, and the differences were on the order of about 20%, which is small compared to odor changes that occur during biosolids cake storage.

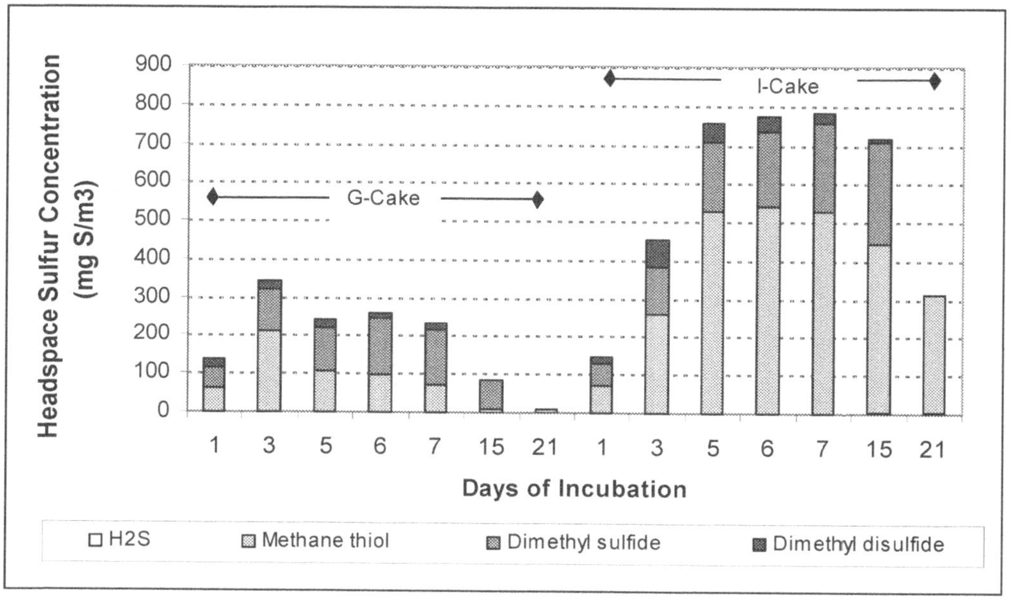

Figure 4-7. Pattern of Volatile Sulfur versus Days of Incubation at WWTP No. 2
(Each Day Sampling a Different Bottle, No Measurements on Day 2 or 4).

A new, airtight bottle was sampled on every day of analysis to prevent headspace gases from escaping through the syringe hole after sampling. This analysis protocol was used to check the ability of cakes to produce and subsequently consume the generated odorous compounds. The headspace bottles are intended to simulate the odor patterns of full-scale biosolids storage in piles or hoppers.

Figure 4-8 indicates that most biosolids cake samples from the test WWTPs release a maximum VSC emission within the first 14 days of storage. WWTP No. 8 had VSC emissions peaking beyond 14 days of storage and had thermophilic digestion. Delayed sulfur emissions were observed around 35 days, potentially due to a shift in microbial population during storage.

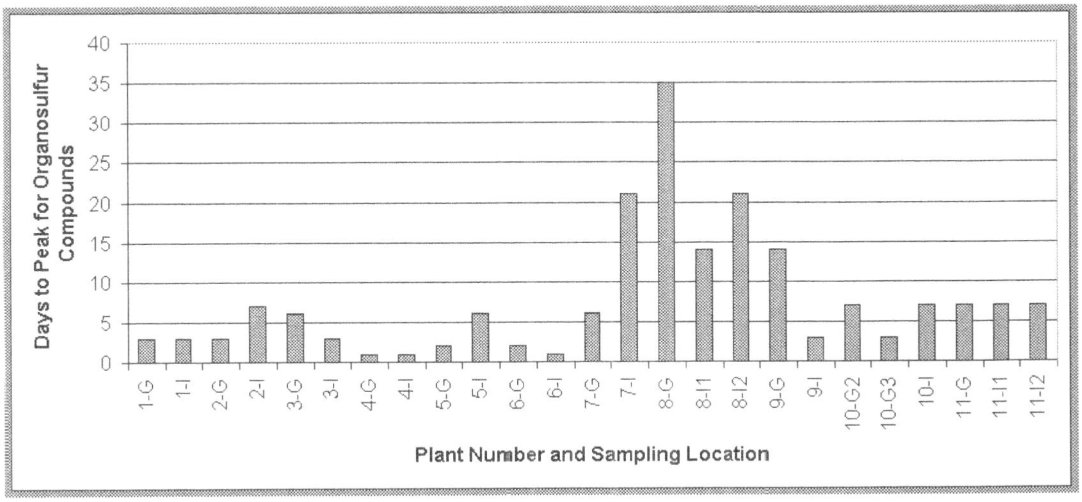

Figure 4-8. Days to Peak for VSC Compounds Measured in Bottle Headspace in Post-Digestion Samples.

Figure 4-9 shows a subsequent VSC decline between the 14th and 21st days. The samples in closed bottles produced concentrated methane, likely the result of methanogenesis, which can utilize methyl groups from methyl-sulfur compounds and trimethylamine to form methane. Consumption of methyl groups in this manner would be reflected in decreased methylated sulfur compounds and increased H_2S in the sample headspace. The H_2S is thought to precipitate primarily as FeS, when there is sufficient Fe in the biosolids sample. Where Fe is at a low concentration or unavailable, an increase in the concentration of H_2S can be expected. This will be discussed further under a subsequent hypothesis.

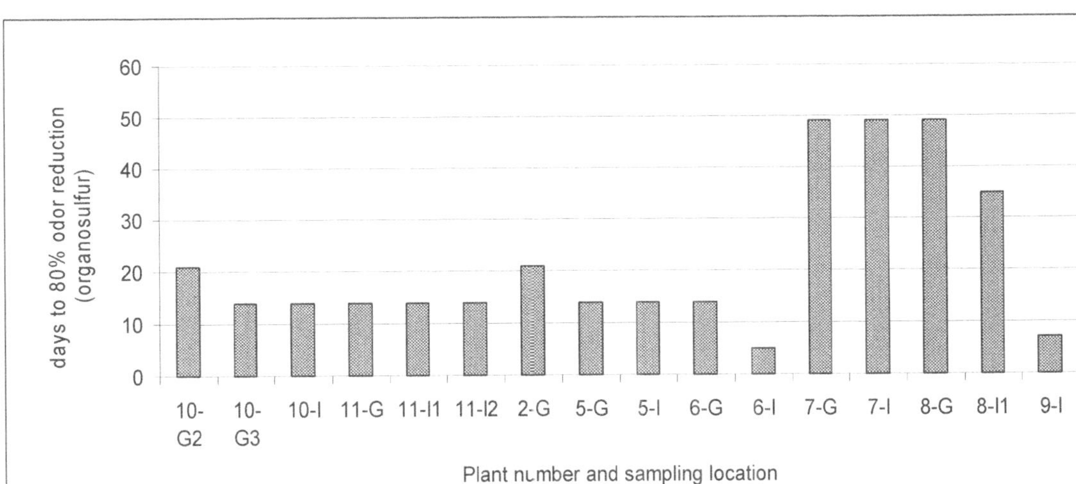

Figure 4-9. Days to Observe VSC Reduction Values Equal or Greater than 80% of Peak in Post-Digestion Samples.

4.2.3 Conclusions

The following conclusions were developed based on the results of this subsection:

- Measuring odorous compounds in the headspace of sample bottles is a practical comparative method for chemical and olfactometric analyses.

- The storage condition in the headspace of the storage bottles was anaerobic, simulating the anaerobic core of full-scale cake piles or storage vessels, since biosolids mixing during storage was not part of the sample bottle handling schedule.
- The headspace of the storage bottles allows the odor consumption of the cakes and the odor production-consumption cycles to be measured. The headspace method is a simple laboratory test to track the changes in sulfur odor that occur during a period of storage before transport. Odor complaints from biosolids often occur during, or as a result of, cake storage.

4.2.4 Recommendations

The following recommendations are based on the results and conclusions of this subsection:

- The bottle headspace method should be considered as part of the odor test protocols for biosolids to make odor measurement simpler and comparable on a plant-by-plant basis.
- For best results, headspace parameters in sample bottles should be controlled, based on:
  - Mass of biosolids and headspace volume of the sample bottle.
  - Oxygen (prevention of unwanted air leakage for anaerobic experiment or wanted addition of pure oxygen for aerobic experiment).
  - Incubation temperature and time, including control during sample shipment.
  - The material of the bottle (leakage, overpressure, and other safety aspects)
- A bench-scale prediction of VSC production and consumption in digested biosolids by anaerobic cake storage should be used for a period that simulates full-scale storage conditions. If biosolids cake cannot be transported from the WWTP within the first day or two, reduction of odors through longer cake storage times might be advised until VSC emissions start to decrease.

4.3 Wastewater Constituents Affecting Biosolids Odors

4.3.1 Role of Protein

4.1.3.1 Hypothesis

The central hypothesis for this research is that bio-available protein is the main substrate for the formation of VSCs associated with odors in biosolids cake. Sulfur-containing amino acids can be degraded to form VSCs. For example, methionine can be degraded to form MT, and cysteine can be degraded to form H_2S (Oho et al., 2000; Persson et al., 1990; Persson, 1992; Higgins et al., 2003). Both MT and H_2S can be methylated to form DMS and MT, respectively (Drotar et al., 1987; Bak et al., 1992; Lomans et al., 2001). In addition, MT can be oxidized to form DMDS (Higgins et al., 2003). The bio-transformations that are mediated by bacteria demonstrate that protein, specifically sulfur-containing amino acids, are the likely substrate for formation of VSC-associated odors. Therefore, greater amounts of bio-available protein should result in greater VSC-associated odors.

4.1.3.2 Results

As part of this study, three different fractions of protein were measured: one soluble and two bound fractions that were extracted from centrifuged pellets of the liquid samples or directly from the cake samples (Subsection 2.3.5). The soluble fraction was measured on filtrate of the liquid biosolids and centrate samples, and the two bound fractions, consisting of a mild and a more rigorous extraction, were measured on the cake samples. The mild extraction used mixing of the sample in PBS at pH 8 for 10 minutes. This fraction is called the "labile fraction," a measure of the easily extracted protein, and is thought to give a representation of the bio-available protein fraction. The more rigorous extraction used 1 Normal (N) NaOH solution mixed with the pellet or cake for two hours. This fraction is termed the "NaOH extract." The soluble protein present in the digester effluent is thought to partition into the cake protein and the centrate protein. The relative fractions of protein in cake and in centrate depend on the amount of binding material, such as Fe and polymer, applied during the dewatering process. As more of the protein is incorporated into the cake, more precursors for odorous compound production will be available after dewatering and during storage.

Although the extractions theoretically represent different protein fractions, the measurement results cannot be summed to determine a "total protein" value or to perform mass balances around the dewatering equipment, since the amount of binding material affected the success of the labile and NaOH extractions as well. In other words, the different fractions of protein as measured here are useful in determining protein availability to support biological activity but are not exact measurements to determine total protein content.

A summary of the different protein fractions from the digester and the cake samples from all the different WWTPs is given in Table 4-4.

Table 4-4. Summary of Protein Concentrations Measured in Digester and Cake Samples from All 11 WWTPs (Ranked from Most Odorous to Lowest According to the Odor Detection Threshold Ranking in Table 4-2).

WWTP No.	Digester Soluble Protein (mg/L)	Digester Labile Protein (mg/L)	Digester Total Protein (mg/L)	Digester Labile Protein (mg/g DS)	Digester NaOH Extract Protein (mg/g DS)	Cake Labile Protein (mg/g DS)	Cake NaOH Extract Protein (mg/g DS)
2-HS	723	433	2682	21.6	133	18.3	177
7	1291	352	4100	16.2	188	24.5	247
1	1029	365	3154	18.1	156	22.6	197
11-HS	402	311	2858	17.8	163	24.2	143
10N	373	542	5486	15.2	154	15.3	155
3	1081	501	6498	22.5	291	27.9	256
8-HS	1814	798	3180	45.9	183	29.9	156
10S	283	387	2955	17.5	134	17.5	134
2	723	433	2682	21.6	133	12.5	121
5-HS	611	436	4998	16.4	188	17.9	206
6	534	484	4067	24.4	205	21.1	180
4-L	596	470	5893	12.9	162	19.8	230
9-PF	1122	570	5021	24.8	218	15.8	204

mg/L = milligrams per liter
mg/g DS = milligrams per gram dry solids

The designations "H" in Table 4-4 stand for "high-solids" centrifuge. Biosolids from all other test WWTPs, except WWTPs No. 4 and No. 9, were processed by low-solids or medium-solids centrifuges. WWTP No. 4 had no dewatering process other than lagoons, and WWTP No. 9 used Plate and Frame (PF) filtration for dewatering. The bound protein content of the cake samples ranged from approximately 12 mg protein/g dry solids (DS) to 30 mg/g for the different WWTPs. The NaOH extract ranged from about 110 to 260 mg protein/g DS. Similarly, the soluble protein in the digesters ranged from 280–1,800 mg/L. The one thermophilic digester (WWTP No. 8) had the highest concentrations of soluble protein.

A good correlation between the amount of labile protein in the headspace bottles and the resultant peak MT constituents was found as shown on Figure 4-10. The R^2 for this regression was 0.61. The data plotted in Figure 4-10 are for the mesophilicly digested cake samples produced by centrifuge dewatering.

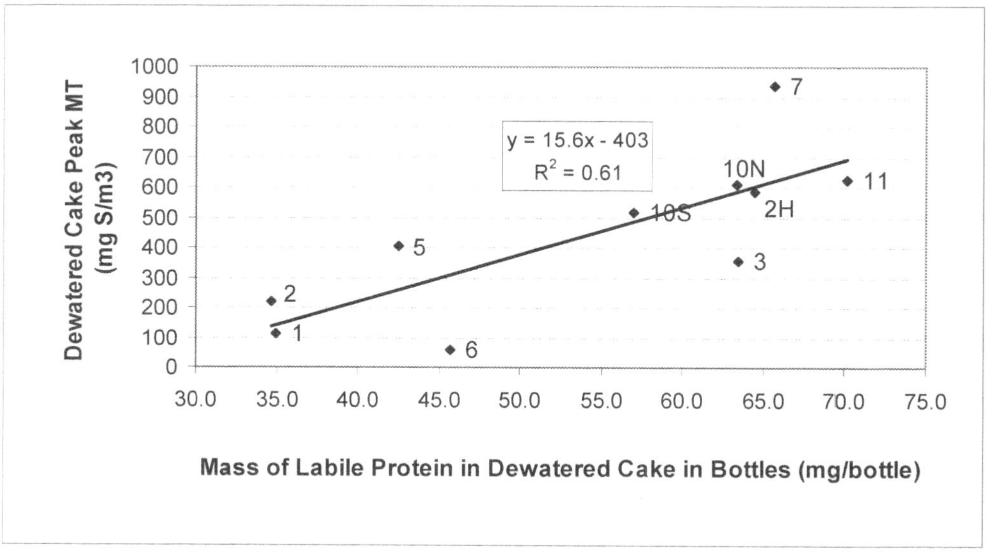

Figure 4-10. Relationship Between Peak MT from Stored Cake and Mass of Labile Protein in Bottles.

A scatter plot of peak MT versus the mass of labile methionine in the headspace bottles is presented in Figure 4-11. The linear correlation for this relationship is excellent, with an R^2 = 0.96, indicating that the main source of MT is methionine degradation.

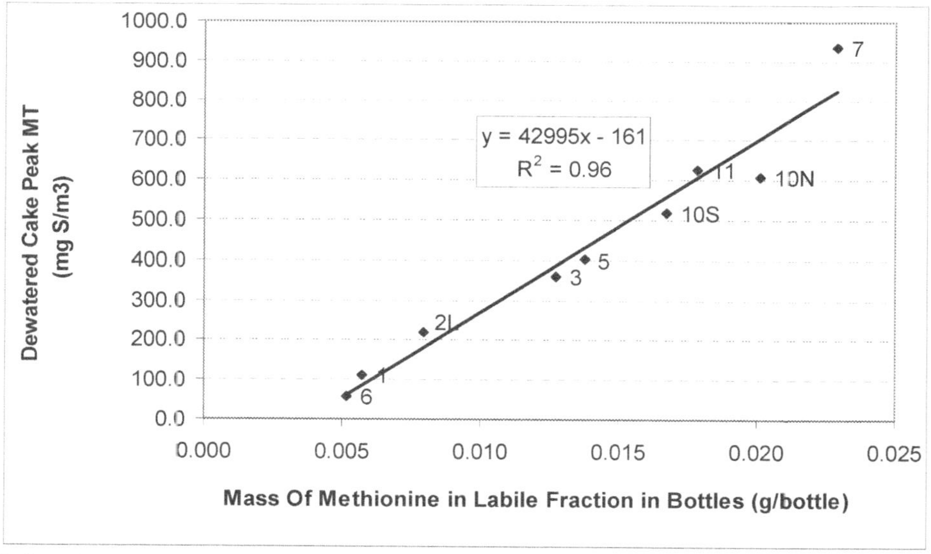

Figure 4-11. Relationship Between Peak MT from Stored Cake and Mass of Labile (Cake Bound) Methionine in Bottles.

The relationship between the labile methionine and odor measurement on Day 6 is also good ($R^2 = 0.72$) as shown on Figure 4-12.

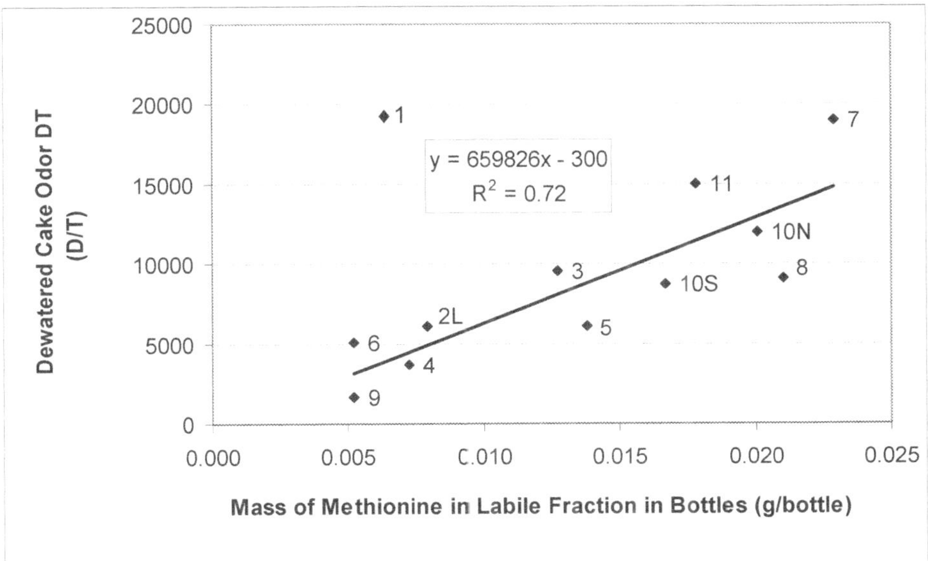

Figure 4-12. Relationship Between Odor DT from Stored Cake and Mass of Methionine in Sample Bottles.

The outlier in this plot is from WWTP No. 1, which had a very high H_2S concentration. Its odors were therefore associated more with H_2S than with the organic VSCs.

4.1.3.3 Discussion

The results showed that significant differences in the protein content of liquid and cake samples exist for the different WWTPs. The protein content in the headspace bottles positively correlated with the peak odors produced by the cake samples. Therefore, greater amounts of protein incorporated in the dewatered cake resulted in greater concentrations of odor-causing compounds. The methionine content in the bottles showed the best correlation with the VSC concentration (as measured by MT). These results support the central hypothesis that protein, and more specifically, the sulfur-containing amino acids that make up the protein, are the main precursors for VSC production from cake samples.

4.1.3.4 Conclusions

- Protein content of liquid digester effluent and cake samples varies significantly from WWTP to WWTP.
- The bound protein fraction and methionine concentration of cake samples correlate very well with the odor and VSC concentration produced from stored cake for this study.
- The results support the hypothesis that protein is the main precursor for VSCs, which are strongly associated with odors generated from stored biosolids cake.

4.1.3.5 Recommendations

The following recommendations are based on the conclusions of this subsection:

- Reducing the bio-available protein concentration in biosolids cake could lead to a reduction in the odors associated with cake storage.
- The mass of bound methionine in biosolids cake samples can be used as an indicator of the odor production potential during storage.
- Further research should be conducted to investigate the impact of protein on nitrogen and sulfur-bearing odorous compounds, testing different types of cake samples processed by different types of biosolids handling equipment.

4.3.2 Role of Enzyme Activity

4.3.2.1 Hypothesis

A fraction of the protein in biosolids is made up of enzymes that are responsible for breaking down protein and producing odorous compounds. The project team hypothesized that enzyme activity may also play a role in producing odor and that greater enzyme activity would be associated with greater odors. If samples had more enzyme activity, this could be an indication of poor digestion performance. As a result, enzyme activity also has potential as a digestion performance indicator.

4.3.2.2 Results

Protein degrading (or proteolytic) enzyme activity was characterized by l-leucine aminopeptidase (LLAP) activity, which is a common enzyme used for this purpose (Teuber and Brodisch, 1977). A summary of the LLAP activity measured in the bound fraction of the cake and digester samples is given in Table 4-5. The LLAP activity was measured on samples the day they arrived at the laboratory for analysis (one day after collection).

Table 4-5. Summary of Proteolytic Enzyme Activity Measured in Digester and Cake Samples from All 11 WWTPs.

WWTP No.	Cake l-leucine aminopeptidase Activity (units/mg Labile Protein)	Cake l-leucine aminopeptidase Activity (units/bottle)	Digester l-leucine aminopeptidase Activity (units/mg Labile Protein)
1	0.3766	193	0.1692
2	0.5502	276	0.2708
2-HS	0.5296	492	0.2708
3	0.7841	351	0.3939
4-L	0.0000	0	0.6977
5-HS	0.2873	172	0.0805
6	0.4068	264	1.0216
7	1.1479	1034	0.1520
8-HS	0.4045	628	0.0000
9-PF	0.7188	418	1.0430
10N	0.2900	265	0.0190
10S	0.0000	0	0.0000
11-HS	0.5628	541	0.4412

HS: High Solids; L: Lagoon; PF: Plate and Frame; N: North plant; S: South plant

Figures 4-13 and 4-14 show the absence of a correlation between LLAP activity and the olfactometric measurements on Day 6 of cake storage for digester and post-digestion cake samples, respectively.

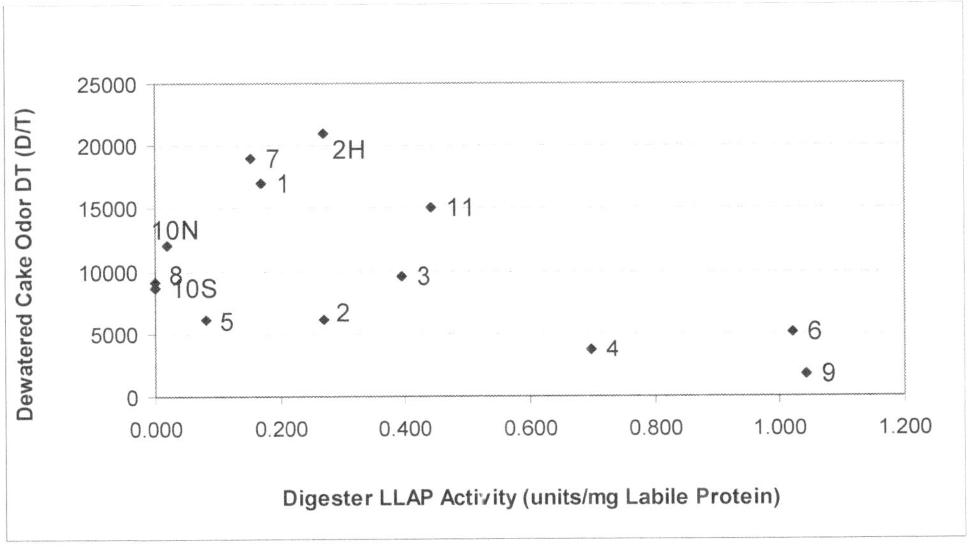

Figure 4-13. Relationship Between Odor Units (D/T) and Initial LLAP Activity of Digester Samples.

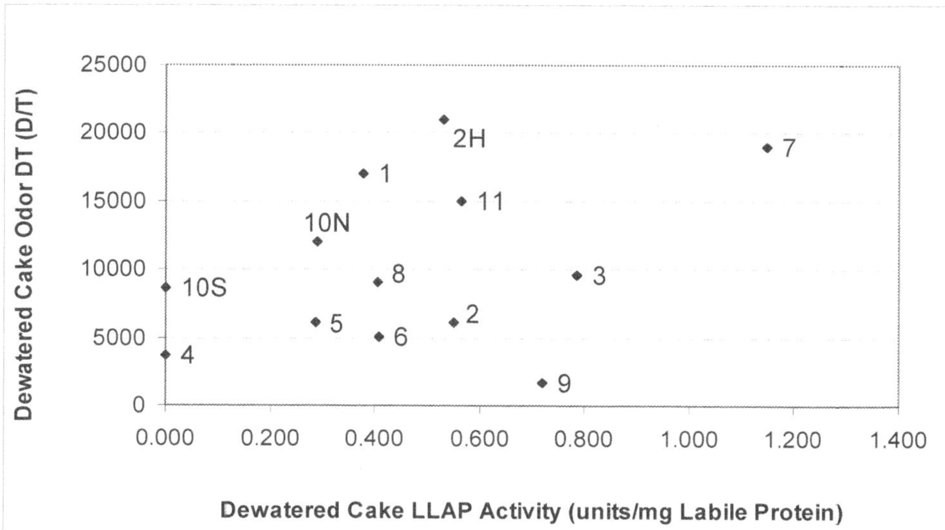

Figure 4-14. Relationship Between Odor Units (D/T) and Initial LLAP Activity of Cake Samples.

The effect of digester SRT on enzyme activity in both the cake and liquid were examined, but no clear relationships were found. For example, Figure 4-15 shows the relationship between digester SRT and cake LLAP activity.

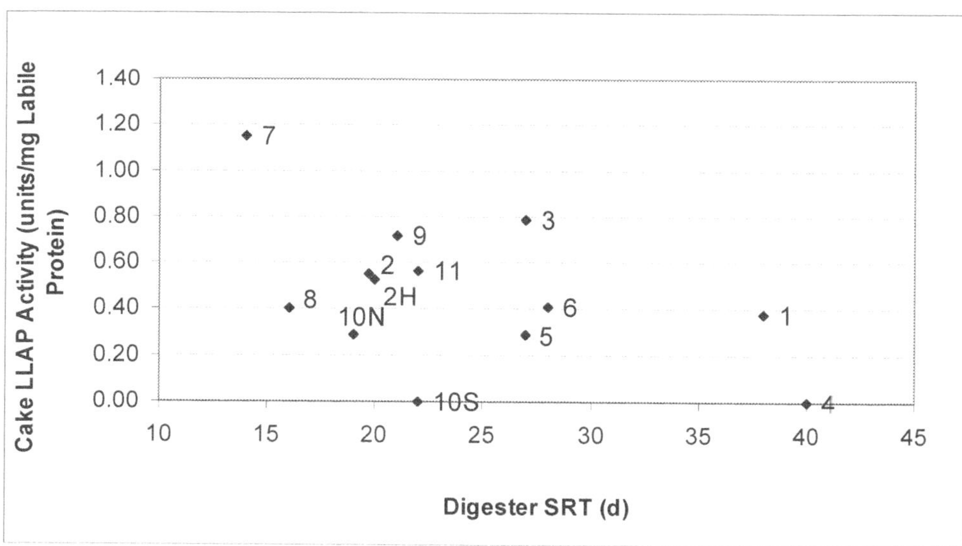

Figure 4-15. Relationship Between Initial LLAP Activity of Cake Samples and Digester SRT.

4.3.2.3 Discussion

Enzyme activity is a measure of the microbial activity of a suspension. Greater enzyme activity indicates that more reactions between the enzyme and its substrate are occurring. The enzyme activity measured in this study was a protein-degrading enzyme. The hypothesis that greater proteolytic degrading activity would create more odors does not seem to be conclusive, since the correlation between cake LLAP and cake odor was poor (Figures 4-13 and 4-14). A number of factors have been shown to affect odor production, and it does not appear that proteolytic activity alone using this particular enzyme assay can be used as a predictor of odor production.

Proteolytic activity has potential as a measure of digester performance in that better digester performance should be associated with lower proteolytic activity. The results from these tests showed that the enzyme activity in the different digester samples varied considerably, from below detection limit (about 0.01 unit/mg protein) to greater than 1.0 unit/mg protein. The relationships between digester performance parameters and digester enzyme activity in the digester, however, did not show any well-defined correlation. For example, no correlation between digester SRT, VS destruction, digester loading rate, and enzyme activity was found.

4.3.2.4 Conclusions

- Protein-degrading enzyme activity measured in cake and digester samples does not correlate well with subsequent odor production from stored cake.
- Protein-degrading enzyme activity does not appear to be a good predictor of digester performance and was not correlated with digester performance parameters, such as SRT, VS destruction, or digester loading rates.

4.3.2.5 Recommendations

- Protein-degrading enzyme activity should not be used as a tool for measuring the odor production potential of a dewatered biosolids sample.

4.3.3 Role of Cations

4.3.3.1 Hypothesis

Cations have been shown to play a key role in the formation of bioflocs, which affect settling and dewatering properties of biosolids (Higgins and Novak, 1997a and b; Sobeck and Higgins, 2002). For example, the divalent-cation bridging theory, which can be used to best explain the role of cations in bioflocculation, states that multivalent cations are important to floc formation, and monovalent cations can cause poor floc formation (Higgins and Novak, 1997a and b; Sobeck and Higgins, 2002). Multivalent cations play an important role in binding biopolymers, mainly protein and polysaccharide, within the floc, which would likely decrease their bioavailability (Higgins and Novak, 1997a). Novak et al. (2002) also suggested that cations Mg, Ca, Fe, and Al associate with different fractions of organic material, mainly protein and lysis products of cellular material bound to biosolids (Figure 4-16). As a result, cations could play a role in odor reduction in that multivalent cations could reduce odors by reducing the bio-available protein content of flocs. In addition, cations such as Fe can play a role in decreasing odor by binding sulfur compounds, such as H_2S and MT, which are associated with odor. However, one other factor that needs to be emphasized is that Fe-bound fraction was suggested to be labile in contrast to the others, meaning that the complexes between Fe and the protein are easily reversible compared to others. Thus, during digestion and post-digestion storage, this fraction, but not the others, can be degraded through hydrolysis and fermentative reactions (Novak et al., 2002).

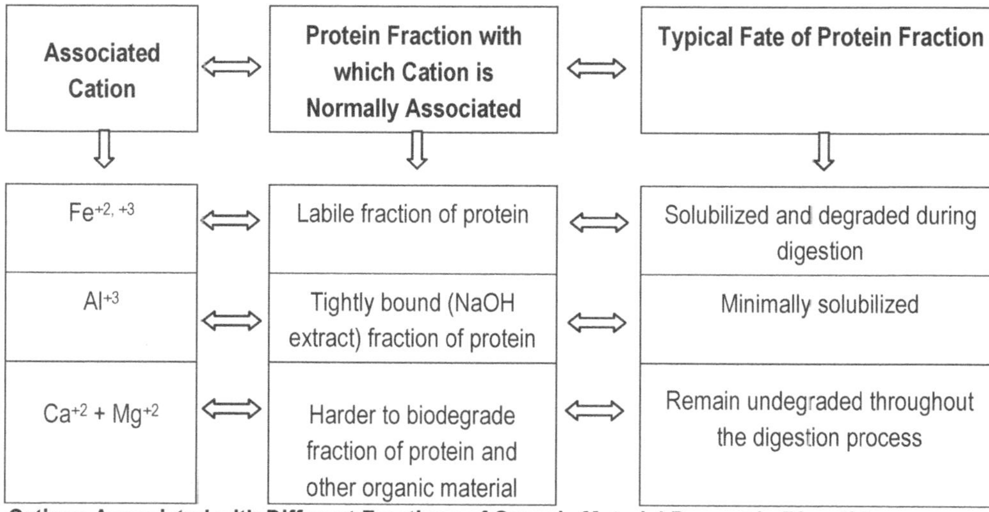

Figure 4-16. Cations Associated with Different Fractions of Organic Material Present in Biosolids.

4.3.3.2 Results

The relationship between the monovalent to divalent cation ratio (M/D ratio) and the soluble digester protein concentration is shown on Figure 4-17. In this study, the M/D ratio is calculated by dividing the sum of the monovalent cation concentrations by the sum of the multivalent cation concentrations, rather than the divalent cations only. All concentrations are expressed in milliequivalents per liter (meq/L). The monovalent cations considered were Na, K, and NH_4, and the multivalent cations considered were Mg, Ca, Fe, and Al. Although there is an apparent trend on Figure 4-17, the correlation was found to be very weak ($R^2 = 0.22$). Thus, the positive trend was not established as a significant correlation.

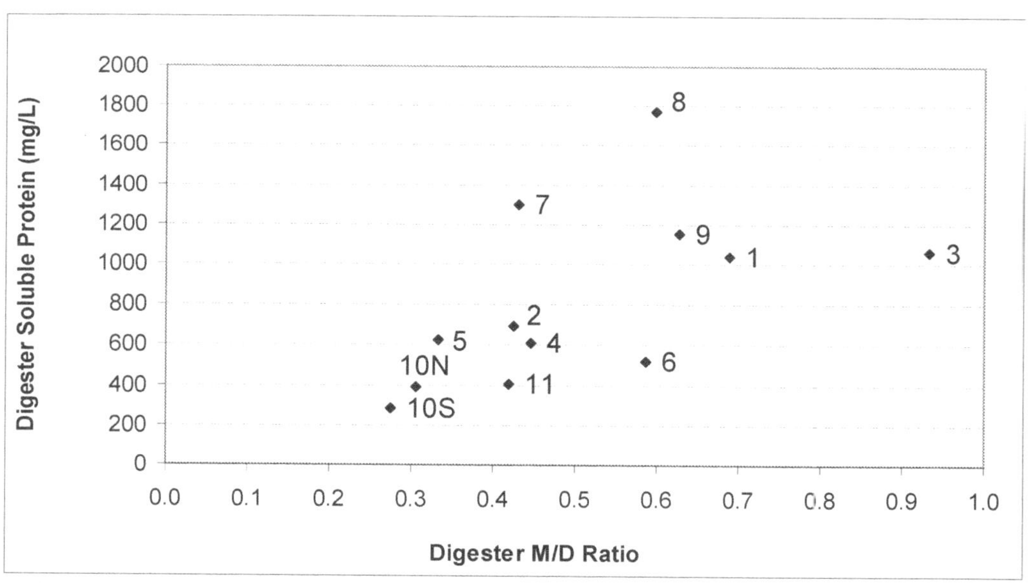

Figure 4-17. Digester Soluble Protein Concentration as Function of M/D Ratio.

Evaluation of the single cations in the digester samples showed no correlation between parameters, except for the total Fe concentration. A plot of the digester total Fe content versus the soluble protein concentration in the digester is shown on Figure 4-18. Digester total Fe versus the labile protein in the dewatered cake is shown on Figures 4-18 and 4-19.

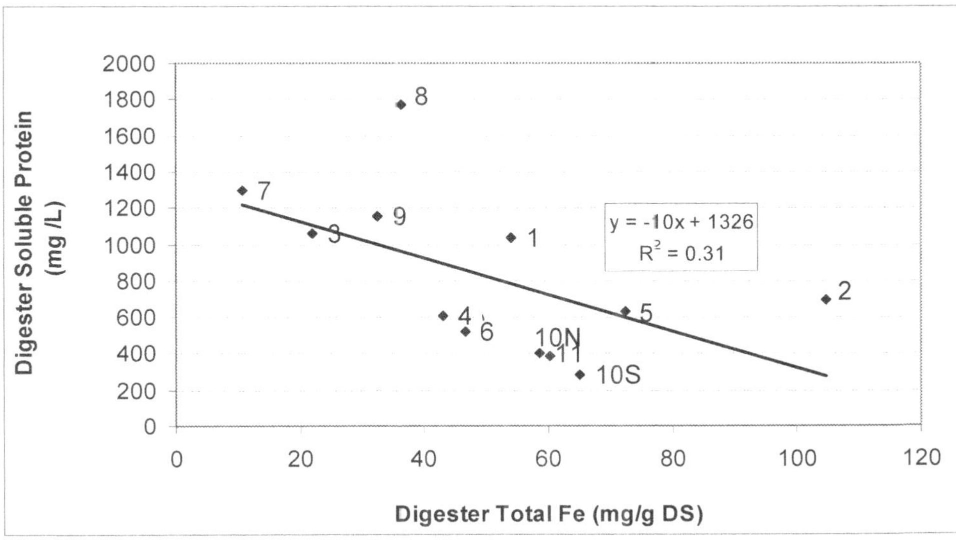

Figure 4-18. Digester Soluble Protein Concentration as Function of Total Fe Content of Digester.

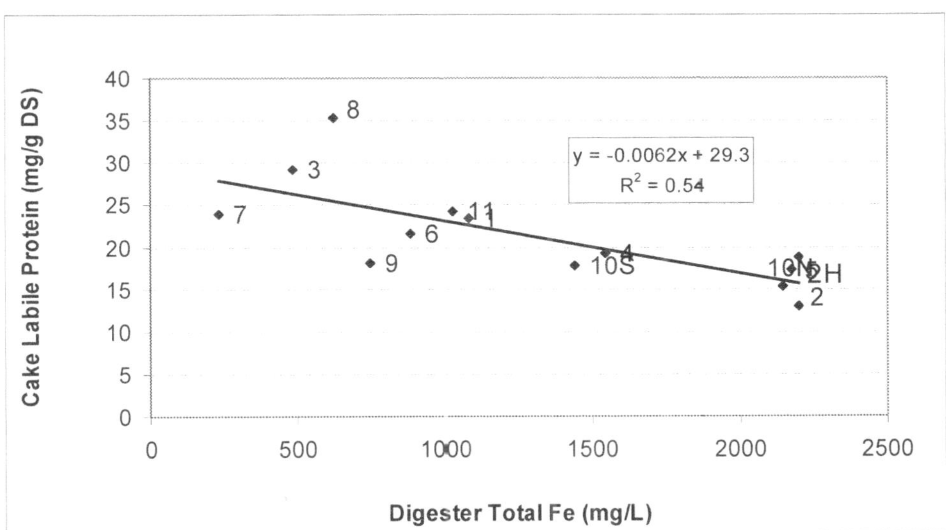

Figure 4-19. Cake Bound Protein Concentration as Function of Total Fe Content of Digester.

4.3.3.3 Discussion

The cation concentration appears to play a role in the floc formation and retention of biopolymer in the floc. The M/D ratio showed a positive relationship with the soluble protein concentration in the digester samples. This fits well with the divalent-cation bridging theory that states increasing amounts of monovalent cations relative to multivalent cations can result in deflocculation with a release of biopolymer (Higgins and Novak, 1997; Sobeck and Higgins,

2002). None of the individual cations correlated well with the protein content, except the Fe concentration, which was negatively correlated with the soluble Fe concentration.

As the Fe concentration increased, generally a decrease in the soluble protein concentration was measured, suggesting that Fe plays a role in binding biopolymer with the floc, which should make it less available for degradation and VSC production. Other researchers have shown that absorption of biomolecules with Fe reduces their degradation rates (Dentel and Gossett, 1982). The protein-binding properties of Fe should have a positive impact on reducing odors as well. Researchers have shown that Fe addition can reduce odor emissions from stored cake (Higgins et al., 2002a; Higgins et al., 2002b). However, although the cation content parameters suggested that there might be a relationship with protein concentration, they did not correlate well with subsequent odor production in the stored cake samples. As indicated, a number of factors can affect odor production, such as protein concentration, dewatering processing, conveyance and storage conditions, and de-complexation of the Fe-bound organic material. Therefore, the ultimate effect of cation concentration on odors is likely complicated by these factors.

4.3.3.4 Conclusions

- Although a direct correlation between the cation parameters and odors was not found, a relationship between cation content and protein concentration is suggested. Therefore, cations should be considered to play some role in affecting the substrate (protein) availability along with other factors for VSC-associated odor production.
- As the M/D ratio increased in the samples, greater amounts of soluble protein were measured in the digester samples, which fits well with the divalent-cation bridging theory for bioflocculation. However, the correlation was not strong enough to confirm.
- The presence of Fe in biosolids indicates a correlation with the soluble protein concentration, in that greater concentrations of Fe appeared to decrease the soluble protein in biosolids cake.
- Cations are one of many factors that affect odor production during cake storage.

4.3.3.5 Recommendations

- Additional research is needed to investigate the effects on odor production of cation addition and reduction of the M/D ratio through Fe addition. However, the mechanisms that lead to release of protein once bound in cake by Fe or cations should also be investigated, since such a release would lead to a delayed odor release during storage or following land application, instead of providing a definite solution to odor generation problems.

4.3.4 Effect of Sulfate

4.3.4.1 Hypothesis

A hypothesis supported by prior studies is that high plant influent sulfate concentration leads to increased odor production by sulfate reduction to form odorous H_2S. Low influent Fe could also increase H_2S, because less H_2S will be precipitated as FeS, but the H_2S concentration in the final cake will depend on the amount of Fe in the digester.

4.3.4.2 Results

The results of influent sulfate, influent iron, digester total iron, dewatered cake peak H_2S (Sample Location G), dewatered cake peak total organo-sulfur (Sample Location G), and dewatered cake olfactometry measurements (DT) (Sample Location G) are presented in Table 4-6.

Table 4-6. Results of Influent Sulfate, Influent Fe, Digester Total Fe, and Dewatered Cake Peak H_2S.

WWTP No.	Influent Sulfate (mg/L)	Influent Total Fe (mg/L)	Digester Total Fe (mg/L)	Cake Peak H_2S (mg S/m3)	Cake Peak Total Organosulfur (mg S/m3)	Cake DT Day 6
1	18	9	1082	1947	169	17000
2	48	12	2200	2	395	6100
3	37	2	492	5	470	9600
4	344	16	1432	18	5	3700
5	9	2	2162	5	673	6100
6	133	11	868	27	144	5100
7	83	1	235	750	1033	19000
8	85	7	635	10	91	9100
9	207	4	751	7	3	1700
10	223	8	2330	13	943	12000
11	49	4	1036	21	848	15000

The stacked bar chart presented on Figure 4-20 illustrates the distribution of the VSC compounds grouped as H_2S, MT, and other organosulfur compounds (a sum of DMS and DMDS) in pre-digestion sampling points, the top of the bar representing the total VSC for each sampling point. The chart shows that H_2S was especially high in the C and D samples of WWTPs No. 1 and No. 7. As can be seen from the distribution of the VSCs, the pre-digestion samples did not contain significant organic VSCs or H_2S, except for WWTPs No.1 and No.7.

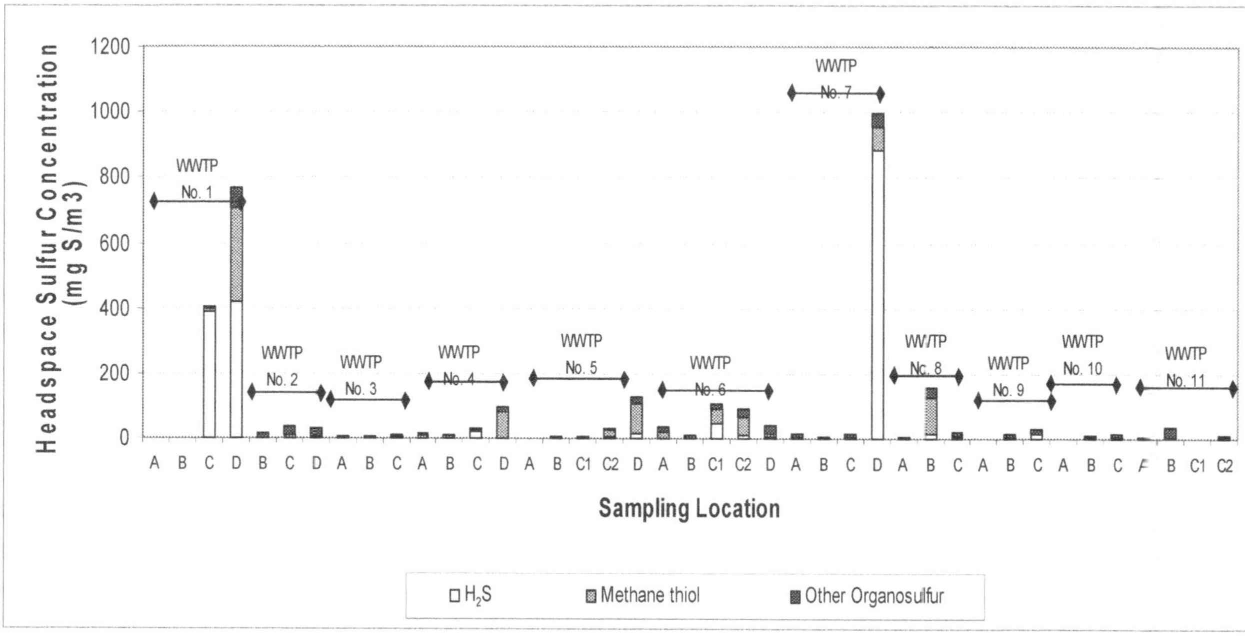

Figure 4-20. Peak Sulfur Species at Sample Locations A Through D (All WWTPs).

Similarly, the stacked bar chart in Figure 4-21 shows that H_2S dominates the sulfur odor for post-digestion cake samples (Sample Locations G and I) of WWTP No. 1 but also contributes significantly to the sulfur odor of WWTP No. 7.

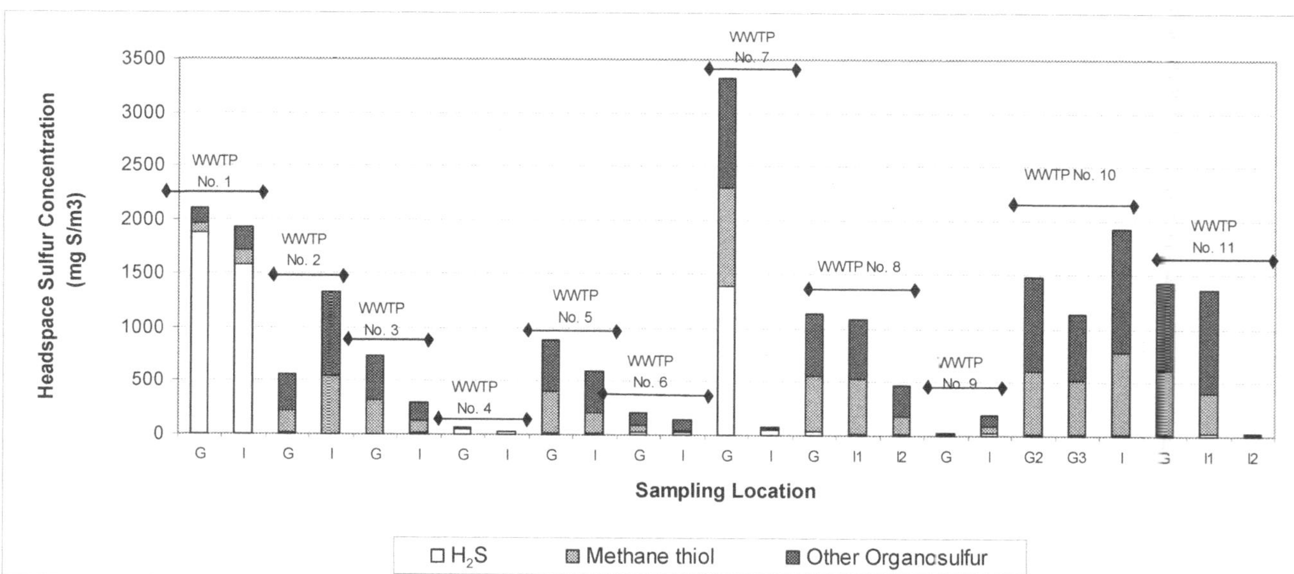

Figure 4-21. Post-Dewatering Peak Sulfur Compounds (All WWTPs).

4.3.4.3 Discussion

The results did not provide conclusive evidence of increased H_2S formation by increased influent sulfate or by low iron. For example, data for WWTPs No. 1 and No. 7 show high H_2S (which also dominates the total sulfur odor), but these WWTPs also have low influent sulfate. Also, the results did not provide evidence for increased organo-sulfur concentrations in conjunction with increased influent sulfate.

Two out of 11 test WWTPs showed significant H_2S concentrations in the headspace gases, which could have been potentially caused by high influent sulfate, insufficient "natural" influent iron, and/or low Fe addition in the collection system or at the WWTP. It is very likely that sulfate reduction and H_2S formation are significant for mesophilic digestion processes. Also, the continuance of "digestion" in the anaerobically stored cake could produce extra H_2S by sulfate reduction. However, elimination of H_2S sulfur in the form of FeS will reduce H_2S-related odors.

Most of the time, the amount of "natural" influent Fe will not be sufficient to eliminate all the H_2S as FeS. But Fe addition as "reactive iron" (soluble iron) can efficiently eliminate H_2S as FeS during digestion or anaerobic cake storage. However, insufficient Fe addition may result in free H_2S remaining in solution, which will be adsorbed by the cake or further formed in the cake by sulfate reduction, thus producing H_2S odor in the stored cake.

4.3.4.4 Conclusions

One means of odor reduction is to ensure that sufficient Fe is left in the cake to precipitate H_2S. One way is to add Fe far in excess. Another way is to use high H_2S as a parameter commanding the addition of more iron. H_2S odor in cake will be low as long as there is sufficient "reactive" Fe left in the cake.

4.3.4.5 Recommendations

The term "reactive iron," a form of iron which can react with odorous H_2S and form non-odorous FeS, should be introduced to odor research,. Future measurement of Fe should be extended from "total" Fe (an essential part of which is FeS, which cannot precipitate additional H_2S) to "reactive" Fe (any form of Fe such as ferric or ferrous Fe which can precipitate H_2S as FeS). Also, other forms of reactive iron, such as Fe loosely bound to biomass or proteins (as "complexes") can reduce H_2S odor through precipitation as FeS.

It follows that Fe addition shows promise in controlling H_2S in biosolids, which was found in significant quantities in two out of 11 WWTPs. A controlled mechanistic study should be performed using Fe addition at one of the test WWTPs that showed high H_2S in its digested biosolids.

4.4 Process Impacts on Biosolids Odor Quality

This subsection discusses the impacts of wastewater treatment and biosolids handling processes on biosolids odor quality.

4.4.1 Pre-Digestion Wastewater Treatment and Solids Handling

4.4.1.1 Hypotheses

WWTPs generally produce two types of solids: primary solids that are collected in primary clarifiers and secondary solids that are wasted from activated sludge units. Primary solids consist mostly of untreated organic and inorganic materials that settle to the bottom of primary clarifiers, whereas secondary solids consist of a mixture of live and dead microorganisms grown on organic materials that pass through the primary clarification process.

Primary and secondary solids are transferred to sludge processing facilities, where they are further treated by a wide variety of processes. Primary solids are usually either thickened in gravity thickeners or pumped directly to digesters, sometimes mixing with secondary solids prior to digestion. Secondary solids are usually thickened prior to digestion, by DAFTs, GBTs, or gravity thickeners. Primary and secondary solids are sometimes mixed prior to digestion and sometimes kept separate until entering the digestion process.

Literature reviewed during Phase I of this project indicated that solids management practices in the pre-digestion processes could influence emissions from biosolids produced by these processes. Hentz (2000a) provided data that showed storage of primary and WAS together for more than 24 hours provided conditions that promote the creation of more odors from biosolids. Bonnin et al. (1990) conducted an odor survey at biosolids processing facilities of 15 WWTPs in West Germany and 12 WWTPs in France. Significant quantities of H_2S and VSCs were emitted from solids thickening facilities, anaerobic digesters, thermal conditioning processes, belt presses, centrifuges, and biosolids storage facilities.

Other researchers (Winter and Duckham, 1999) examined the effect of adding WAS at differing proportions and sludge age to the anaerobic digesters. Although no trend was specifically proven, the data implied that WAS could increase odor emissions in the final product. Devai and DeLaune (2000) showed that the emission of H_2S and organic sulfur compounds from thickened biosolids was influenced by the ORP of the biosolids. H_2S emissions were highest at –220 mV and undetectable at +340 mV. Considerable organic sulfur emissions were measured at moderate ORP ranges between –160 mV to +60 mV.

The WWTPs tested in this project employ many of the pre-digestion processes described above. Thus, the data gathered in this project allow the limited observations found in the literature to be further investigated. Specifically, the data in this project were evaluated for the following hypotheses:

- *Hypothesis Number 1* - Mixing of primary and secondary solids in pre-digestion processes can increase odor emissions from the pre-digestion processes and digested biosolids.
- *Hypothesis Number 2* - WAS has a higher potential than primary solids to produce more odorous biosolids following anaerobic digestion.
- *Hypothesis Number 3* - Increased storage time for primary solids can lead to septicity, which can cause increased odor emissions from pre-digestion processes and digested biosolids.

- *Hypothesis Number 4* - Longer WAS SRT in the activated sludge process can produce less odorous biosolids after digestion.

4.4.1.2 Results

Peak sulfur emissions from solids at Sample Locations A through D are shown on Figure 4-22 for each WWTP. In general, these sample locations represent the following solids and processes:

- Sample Location A = WAS from the activated sludge process.
- Sample Location B = WAS after thickening and storage (if provided).
- Sample Location C = Primary solids after storage (if provided).
- Sample Location D = Combined WAS and primary solids just before addition to the anaerobic digesters.

Exceptions to these designations are at WWTPs No. 2, No. 4, No. 7, and No. 8, where the following types of solids were collected:

- At WWTP No. 2, Sample Location C was thickened WAS and Sample Location B was primary solids.
- At WWTP No. 4, Sample Location D was trucked waste that was added directly to the anaerobic digester.
- At WWTP No. 7, Sample Location A was thickened WAS and Sample Location B was primary solids. At WWTP No. 8, Sample Location B was thickened WAS and Sample Location C was primary solids.

The H_2S and VSC emissions for the solids from all of these sample locations were very short-lived. H_2S and VSC concentrations in the headspace of the sample bottles generally peaked within the first 24 hours of storage and decreased to very low concentrations after 24 hours. This suggests the H_2S and VSCs were absorbed back into the cake matrix and consumed by biological activity. It is noteworthy to mention that the bottles contained less VS for these sampling locations compared to the post-digestion sampling points, since each bottle contained the same mass of samples, which included the water content of the samples along with the dry solids.

As was shown in Figure 4-21, H_2S emissions were by far the highest at Sample Location D in WWTPs No. 1 and No. 7, which involved combined storage of WAS and primary solids prior to digestion. Table 4-7 also shows that WWTPs No. 1 and No. 7 had the lowest ORP and pH readings at Sample Location D, indicating highly reduced conditions strongly favoring H_2S generation. This further indicates that combining WAS and primary solids can lead to septic conditions, which can result in odor generation.

Primary solids (Sample Location B) from WWTP No. 1 also produced significant H_2S emissions, which continued through the anaerobic digestion and dewatering processes, and created cakes that had the highest sulfur emissions. However, the dewatered cake from WWTP No. 7 also emitted a significant quantity of MT.

Table 4-7. Influent Sulfate ORP, Influent Fe, Digester Total Fe, and Dewatered Cake Peak $H^{2}S$.

	Thickened WAS			Primary Sludge				Combined Sludge				Gas		Digested Sludge	
WWTP	pH	Site	ORP	H_2S	PH	Site	ORP	H_2S	PH	Site	ORP	NH_3	H_2S	pH	ORP
1	6.3	B	-177	350	5.5	C	-192	70	5.5	D	-196	60	30	7.1	-293
2	6.8	C	-179	0.49	6.4	B	-160	0.19	6.4	D	-169	3	0.01	7.4	-241
3	6.6	B	-68	1.45	6.0	C	-77					20	0.18	7.0	-210
4	6.5	B	-65	23.0	6.11	C	-63					75	6	7.0	-213
5	6.4	B	-113	130	6.7	C1	-228	0.03	6.4	D	-180	53	0.023	7.0	-240
				14.0	5.9	C2	-265								
6								7.7	5.9	A	-183	28	0.21	7.0	-268
7	7.1	A	-136	23.0	7.4	B	-182	10.7	5.6	D	-200	18	7	7.0	-307
8	6.8	B	-174	0.12	6.1	C	-116					33	0.021	7.6	-125
9	7.0	B	Too thick	5.3	5.8	C	24					50	0.11	7.2	33
10	6.8	B	-109	1.7	5.6	C	-50					20	0.18	7.1	-196
11	6.5	B	-138	10.0	5.7	C2	-126					14	0.11	7.1	-175

The stacked bar chart in Figure 4-22 shows H_2S and VSC emissions after the H_2S values from Sample Locations C and D at WWTP No. 1 and Sample Location D at WWTP No.7 are removed. Solids from these three sample locations emitted mostly H_2S. Removing these three sample locations allows all other sample locations to be compared. As can be seen on Figure 4-22, most of the other sample locations emitted mostly organosulfur. In particular, MT was the most common VSC emitted. WWTP Sampling Locations 4C, 5D, 6C, 6D, and 8B emitted the highest concentrations of VSCs. All these sampling locations represent either primary solids (Sample Location C) or a combination of WAS and primary solids (Sample Location D).

The stacked bar chart presented on Figure 4-23 shows that MT was the predominant VSC in the dewatered cakes from these WWTPs. However, a comparison of Figures 4-22 and 4-23 shows that MT emissions in the cakes were not proportional to the MT emissions in pre-digestion processes.

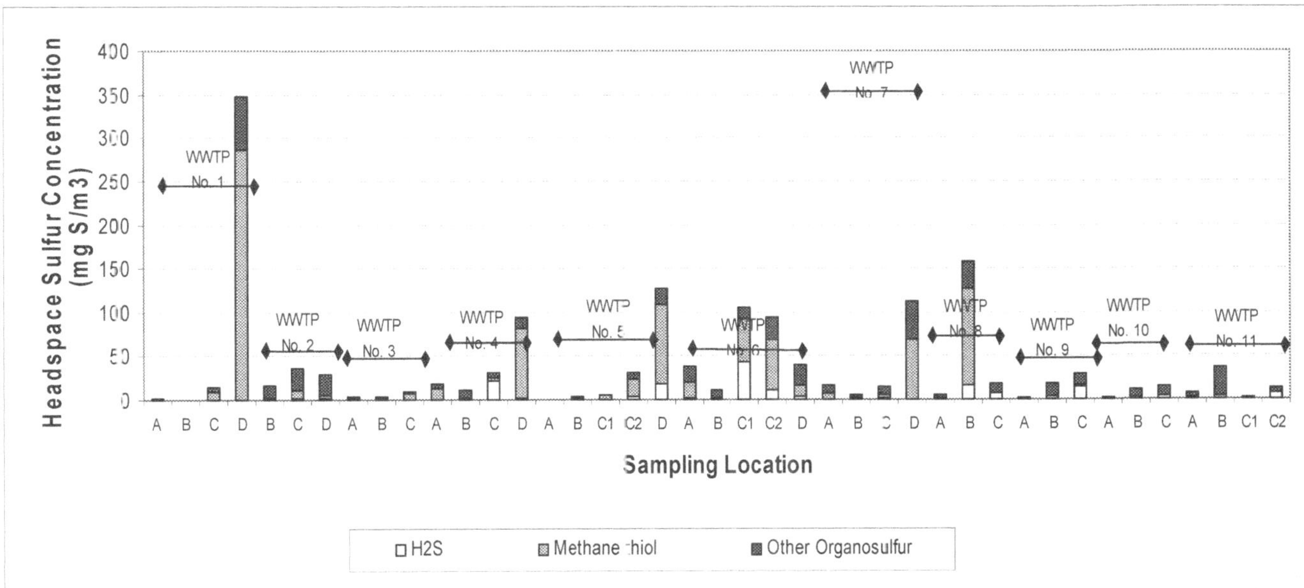

Figure 4-22. Pre-Dewatering Peak Sulfur Compounds
(Not Including H_2S values for WWTP No. 1 Sample Locations C and D, or WWTP No. 7 Sample Location D).

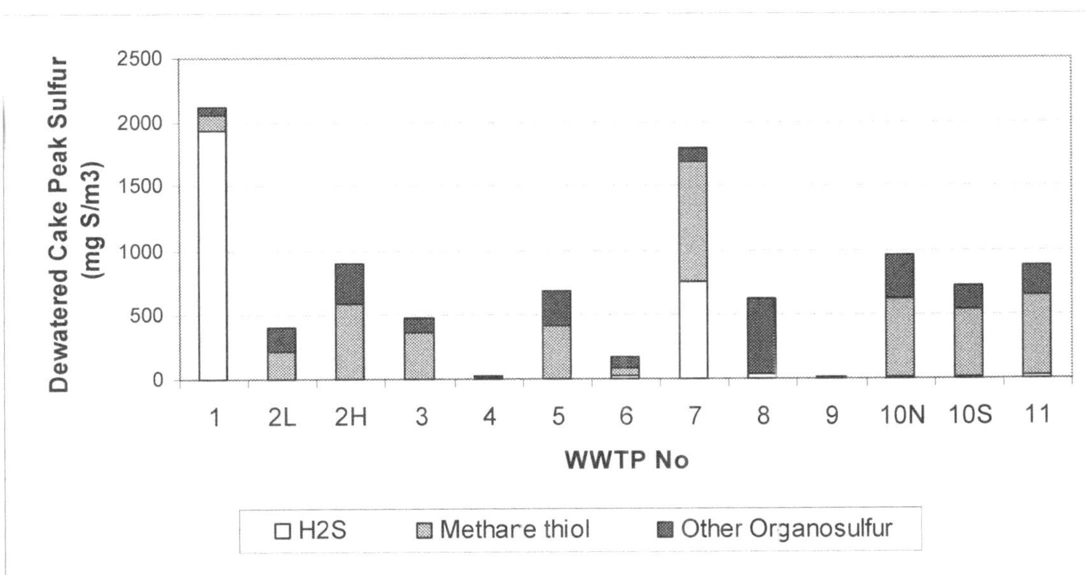

Figure 4-23. Dewatered Cake Peak Sulfur Compounds (All WWTPs).

Figure 4-24 compares peak sulfur emissions from dewatered cakes against the percentage of WAS in the cake. As indicated above, emissions from cakes at WWTPs No. 1 and No. 7 were primarily H_2S. Emissions from the remaining WWTPs were primarily MT. As shown on Figure 4-24, there was a very good correlation (R^2=0.84) between organosulfur emissions from samples obtained from WWTPs that employ mesophilic digestion followed by centrifugation and the percentage of WAS in digester feed. Although it was a mesophilic digestion and centrifuge dewatering plant, WWTP No. 5 data indicated that cakes containing mostly WAS can have higher odor and VSC emissions than cakes that contain mostly primary solids.

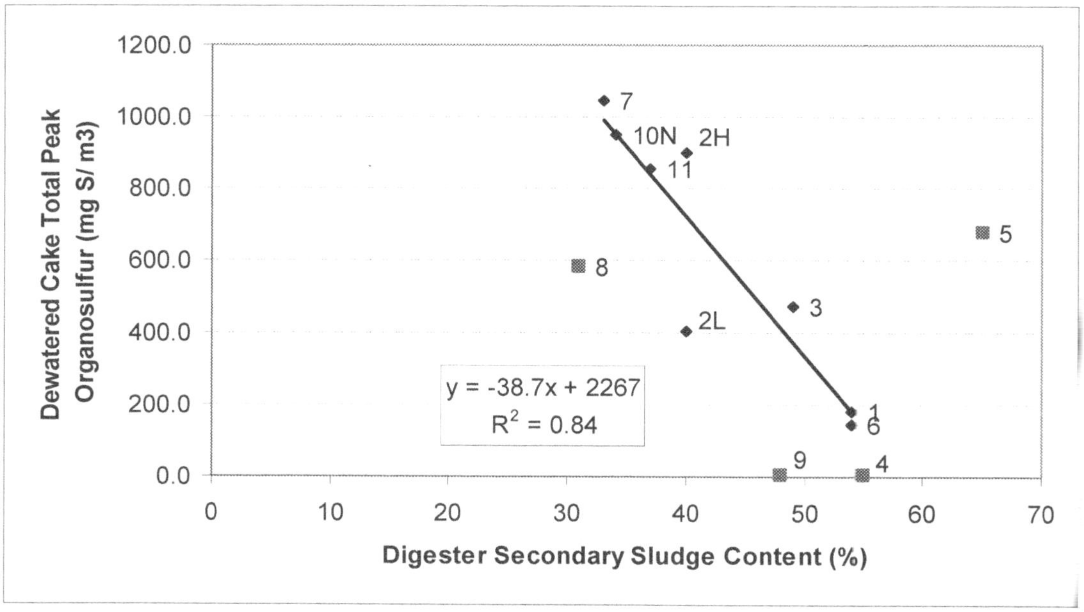

Figure 4-24. Peak Organosulfur Emissions from Dewatered Cakes and Percentage of WAS in Digester Feed.

Figure 4-25 illustrates the correlation between peak sulfur emissions from dewatered biosolids and total detention time for primary solids. As shown, there was no relationship between sulfur emissions and primary solids detention time. Thus, primary solids detention time does not appear to affect the odor quality of the digested cake.

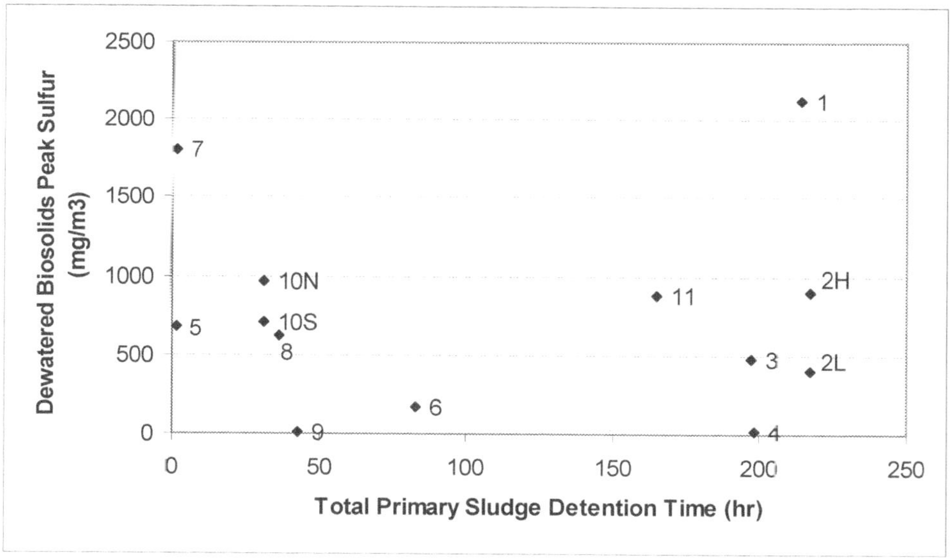

Figure 4-25. Dewatered Cake Peak Sulfur Versus Total Primary Sludge Detention Time (All WWTPs)

Figure 4-26 provides a plot of peak organosulfur emissions from dewatered cakes against total WAS detention time, which included SRT in the activated sludge process and WAS storage time in solids processing facilities prior to digestion. In general, secondary treatment SRTs in all test WWTPs were relatively low (all less than seven days). As shown on Figure 4-26, there was a general trend toward lower organosulfur emissions as the WAS detention time increased. However, the data were insufficient to substantiate a definite relationship at this time.

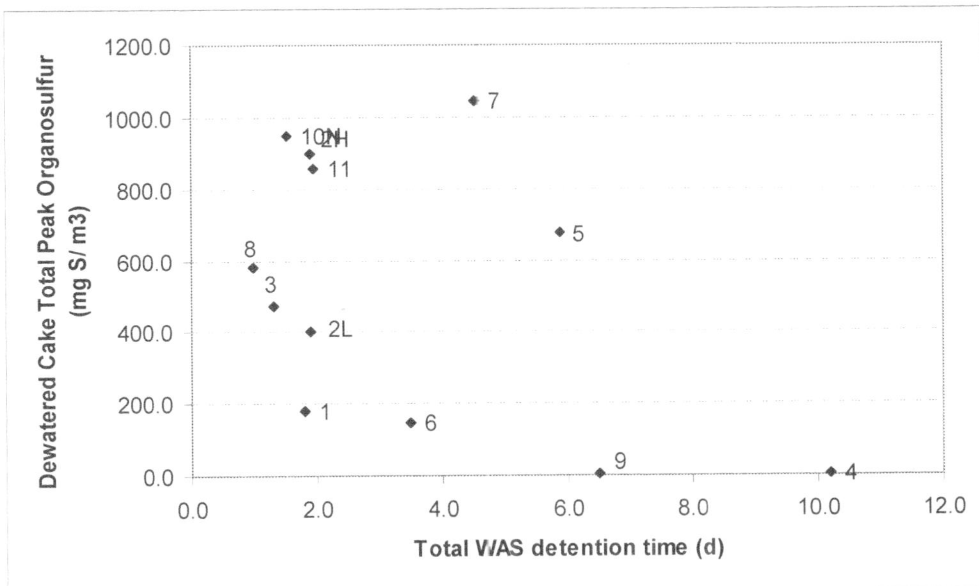

Figure 4-26. Dewatered Cake Peak Organosulfur Versus Total WAS Detention Time (All WWTPs).

4.4.1.3 Conclusions

Based on the data collected, the following conclusions can be drawn regarding the four hypotheses postulated in this subsection:

- *Hypothesis Number 1*: It is likely that mixing primary and secondary solids in pre-digestion processes can increase odor emissions from the pre-digestion processes and digested biosolids. Mixing of WAS and primary solids appears to accelerate the hydrolysis and fermentation processes, reducing the pH and ORP of the biosolids, thereby making more protein available as a substrate for odor-forming biological activity.
- *Hypothesis Number 2*: WAS does not appear to have a higher potential than primary solids to produce more odorous biosolids after anaerobic digestion. To the contrary, a possible, positive correlation between organosulfur emissions from post-digestion biosolids and the percentage of WAS in digester feed is indicated in Figure 4-24, but data are insufficient to draw a conclusion.
- *Hypothesis Number 3*: Increased storage time for primary solids does not appear to increase the odor emissions from pre-digestion processes or digested biosolids.
- *Hypothesis Number 4*: Longer WAS SRT in the activated sludge process may or may not produce less odorous biosolids after digestion. While some of the data indicated a slight trend toward lower organosulfur emissions from dewatered biosolids as the WAS

detention time increases, data collected thus far are insufficient to substantiate a definite relationship.

4.4.1.4 Recommendations

Controlled studies should be conducted to establish whether mixing and storing primary sludge and WAS influences the degradation of biosolids during digestion, thereby affecting post-digestion biosolids odors. Since the results on primary sludge detention times were not conclusive, laboratory studies should be conducted to determine the influence of mixing ratios, detention time, and protein degradation on biosolids odor quality.

4.4.2 Impacts of Digestion on Biosolids Odor Generation

4.4.2.1 Hypotheses

Anaerobic digestion of biosolids produced at WWTPs is one of the most widely used and cost-effective means of solids stabilization prior to further treatment and/or disposal. Anaerobic digestion is generally preferred over aerobic digestion at larger treatment facilities (greater than 10 mgd capacity), due to its lower operation costs and its general effectiveness in meeting regulatory requirements regarding VS destruction and pathogen destruction.

Anaerobic digestion is the self-destruction and consumption of organic material present in the waste primary and secondary biomass, under anaerobic conditions (absence of external electron acceptors such as O_2, NO_3^- and NO_2^-, SO_4^{-3}, etc.) through fermentative reactions. The bacterial populations present in biosolids are responsible for the hydrolysis, breakdown, and stabilization of higher forms of organic material, such as carbohydrates, proteins and lipids. These complex organic compounds originate either from influent wastewater or from biological treatment processes at the WWTP. In either case, destruction of the complex organics requires extracellular hydrolysis of the material into less complex soluble organics, before intracellular enzymes can provide further breakdown and acidogenesis. Once low molecular weight organics are formed and more oxidized terminal electron acceptors are exhausted, methanogenesis takes over for further stabilization of the organic carbon under favorable conditions Methane is the desired end product of this fermentation process, because it can be combusted to generate energy.

The delicate balance between different bacterial groups makes anaerobic digestion a complicated process, requiring close attention from WWTP personnel. At times when this balance is disturbed due to changes in the digester feed or operation, methanogenesis can be inhibited and recovery can be slow, especially considering that methanogens are slow growers. The balance between methanogens and sulfate reducers may become important, if the digester feed contains high sulfate concentrations that promote the growth of sulfate reducers that act as an alternative electron acceptors, affecting methanogenesis. Temperature, pH, solids content and retention time, and hydraulic retention time are also important factors affecting digester performance (WEF MOP 8, 1998). Prior studies have shown that odorous compounds containing sulfur and nitrogen originate from anaerobic degradation of proteins and amino acids (Bonnin et al., 1990; McGrath and Lambert, 2000; Beaman and Winter, 2000; Einarsen et al., 2000; Islam et al., 1998). Thus, based on collective experience of the project team and results of prior research on the subject, the following hypotheses were developed as part of this study:

Hypothesis Number 1: "Good digestion" (usually defined by VS destruction and methane gas production) leads to minimized odors in digested biosolids, and conversely, poor digester performance exacerbates odors in the digested biosolids.

Hypothesis Number 2: Thermophilic digestion can create a different time pattern of odor release and a different odor quality than mesophilicly digested biosolids.

4.4.2.2 Results

In order to further examine the concept of anaerobic digestion and its impact on biosolids odor generation, the project team evaluated digester operation data collected from the 11 test WWTPs. Prior studies and some of the data from this study indicate that longer digestion SRT results in lower odor and sulfur emissions from liquid digested biosolids, when measured immediately downstream of digestion, as illustrated in stacked bar chart presented in Figure 4-27. However, when the project team investigated the odor potential of the digested biosolids in terms of the traditional digester performance indicators, such as VS reduction, the current study did not confirm this hypothesis. Digested biosolids odor quality prior to dewatering is not an indicator of dewatered biosolids odor quality. Digested biosolids quality and its changes during dewatering require further study, evaluating parameters such as:

- Digester effluent VFAs
- SRT
- RBA
- VS reduction
- NH_3 content in digester off-gas
- Methane content in digester off-gas

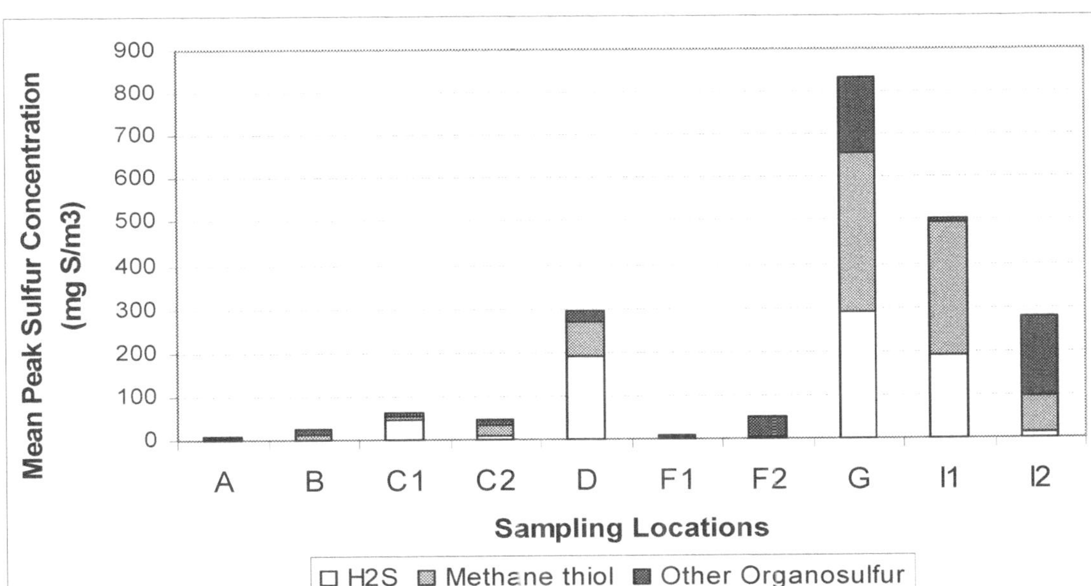

Figure 4-27. Sulfur Distribution at Different Points of Treatment Train Measured on Days 1, 3, 5, and 7 of Sample Storage (All WWTPs).

Digester Effluent VFAs: A high concentration of acetic acid (a short-chain VFA) in digested solids has historically been an indicator of poor digester performance, which is often thought of as a precursor to biosolids odors. However, as shown in Figure 4-28, no clear relationship can be drawn from the data in this study. Even when the apparent outlier WWTPs No. 1 and No. 9 were omitted, the R^2 was 0.40. Figure 4-28 illustrates the variability in the dewatered cake olfactometry measurements (detection thresholds) when plotted against digester effluent acetic acid concentrations. All but two WWTPs had acetic acid concentrations below 200 mg/L, yet the odor (DT) in biosolids cake samples varied between 3,700 and 21,000 D/T. When the same correlation was plotted for WWTPs with medium-solids centrifuges only (not shown), the variation narrowed to between 5,100 and 12,000 D/T in biosolids cake when acetic acid concentration was below 200. WWTP No. 1 had the highest acetic acid concentration in its digester effluent (994 mg/L) and showed a relatively high odor level of 17,000 D/T. The high-solids centrifuge plants with better digester efficiencies (lower acetic acid levels in the digester effluent) still showed relatively high odor levels in the dewatered biosolids, indicating that other factors beyond digestion influence biosolids odor quality.

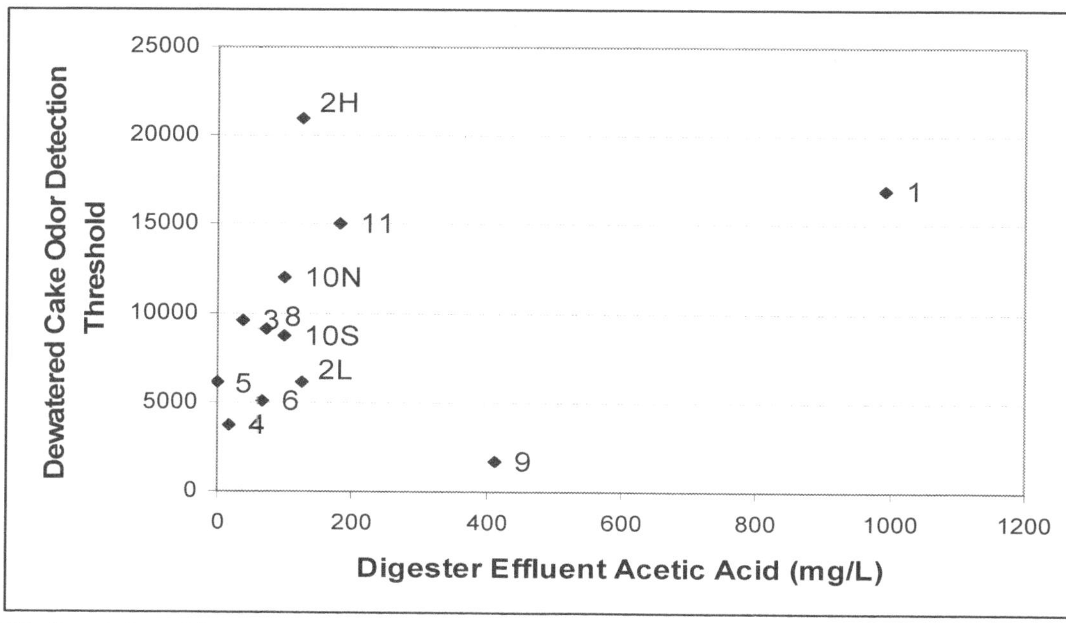

Figure 4-28. Correlation of Dewatered Cake Olfactometry Measurements (DT) with Digester Effluent Acetic Acid (All WWTPs).

Digester SRT: Figure 4-29, plotted for all WWTPs, does not indicate a relationship between digester SRT and cake odors ($R^2 = 0.06$). Data points in this figure included WWTPs without centrifuge dewatering (No. 4 and No. 9) and with thermophilic digestion (No. 8). WWTP No. 1 appears to be an outlier in Figure 4-28 due to very high H_2S concentrations in most biosolids samples. However, for low-solids centrifuge plants, no correlation was found.

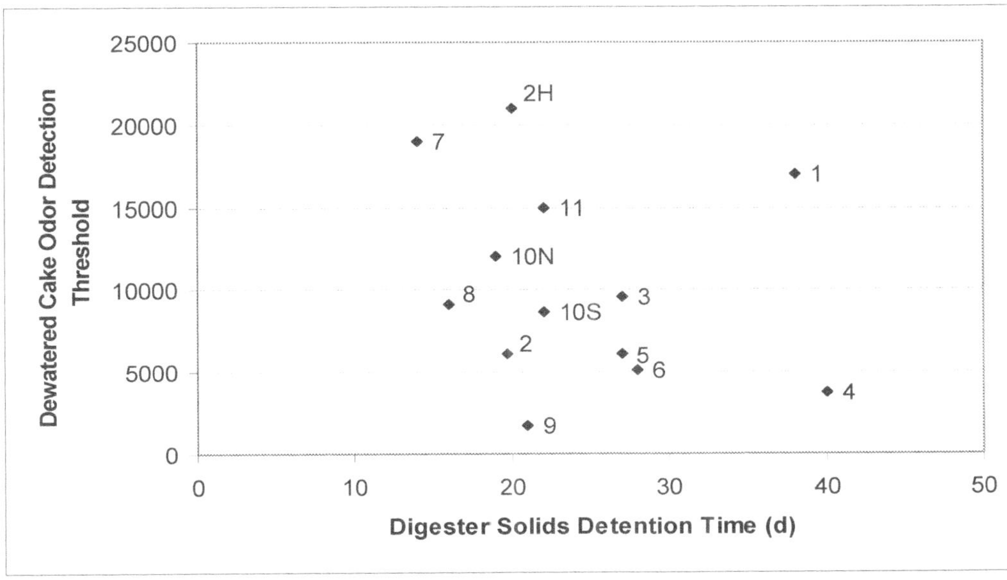

Figure 4-29. Impact of Digester Solids Detention Time on Dewatered Cake Olfactometry Measurements (DT).

Figure 4-30 indicates a fairly strong relationship ($R^2 = 0.62$) between digester SRT and peak organosulfur emissions measured on biosolids samples collected from WWTPs mesophilic digestion followed by centrifuge dewatering. Longer SRT values appear to result in lower peak organosulfur values in this restricted case. The olfactometry data did not show as good a correlation with digester SRT as the organosulfur compounds did, presumably because of the presence of odorous compounds other than those that were measured in this study. Also, results for WWTPs No. 2 and No. 9 indicate that the type of biosolids dewatering process has an impact on cake odors, a factor that needs to be further examined among WWTPs employing similar types of dewatering equipment.

Volatile Solids (VS): Based on prior research, the project team believed that digester feed VS content might play a role in the production of biosolids odors and VSC release. The relationship of VS concentration in the digester feed to dewatered biosolids odors found in this study is illustrated on Figure 4-31. There is no correlation between the two parameters ($R^2 = 0.001$).

The project team also believed based on prior research that higher VS destruction in the digester should have a beneficial impact on digested biosolids quality and dewatered biosolids odors. To investigate this potential relationship, dewatered biosolids odor levels were plotted against digester VS destruction, calculated from WWTP data (Figure 4-32). Odor levels from dewatered biosolids varied within a wide range (DT between 85,000 and 1,050,000 D/T) for a 42-67% VS-destruction range. No correlation was apparent from either this relationship or a plot of dewatered cake VS destruction and peak organosulfur (Figure 4-33), both at a 0.02 R^2 value.

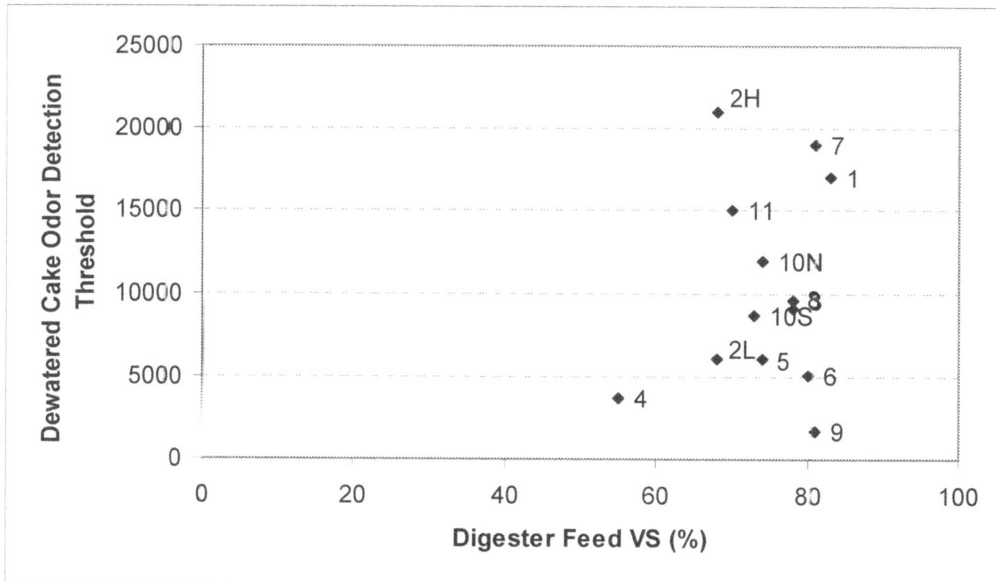

Figure 4-30. Impact of Digester Solids Detention Time on Dewatered Cake Headspace Peak Organosulfur Concentration (WWTPs No. 4 and No. 9 do not have centrifuge dewatering; WWTP No. 8 has thermophilic digestion).

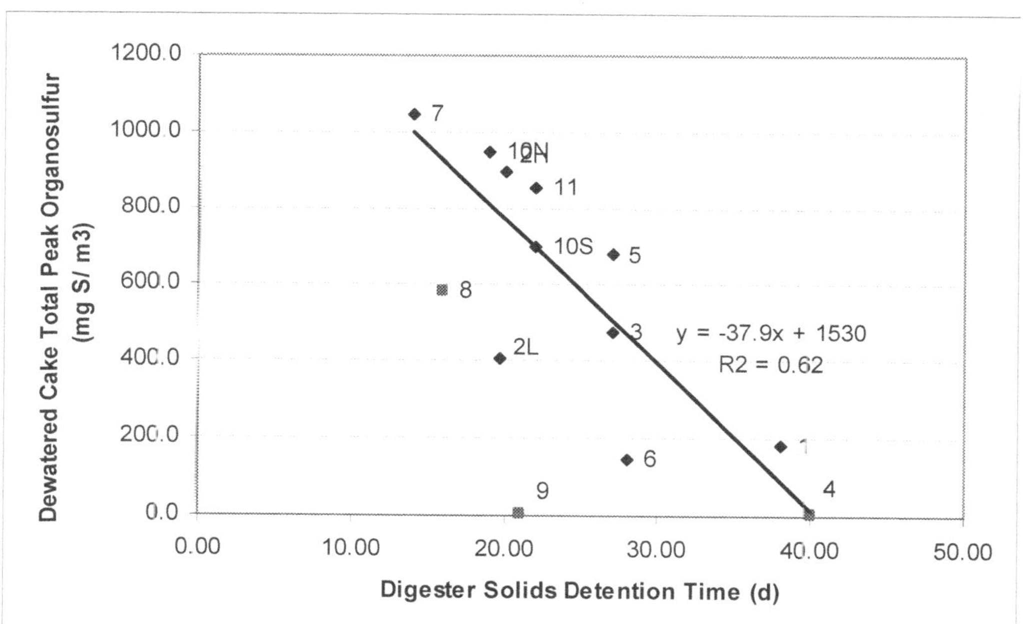

Figure 4-31. Impact of Digester Feed VS Content on Dewatered Biosolids Olfactometry Measurements (DT).

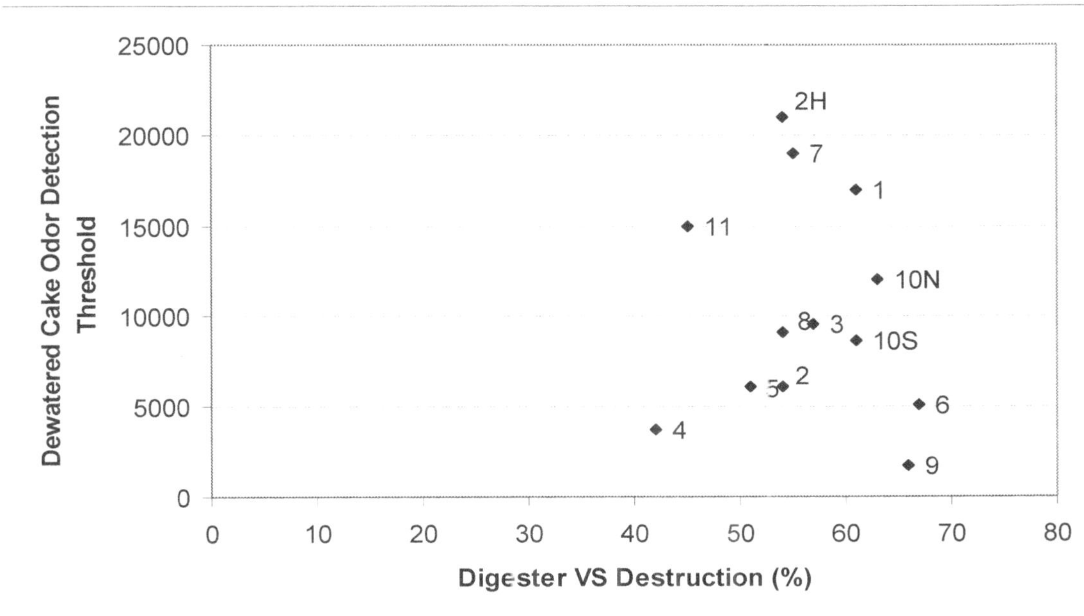

Figure 4-32. Impact of Digester VS Destruction on Dewatered Biosolids Olfactometry Measurements (DT).

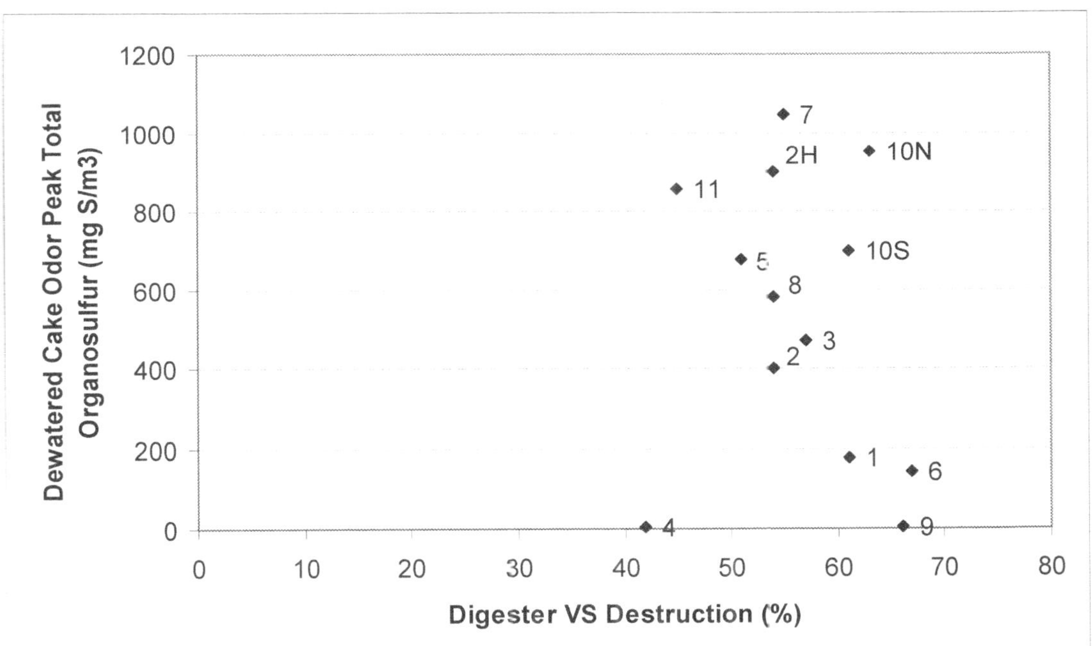

Figure 4-33. Impact of Digester VS Destruction on Dewatered Biosolids Headspace Peak Organosulfur Concentration.

Evaluation of the data showed that VS destruction in anaerobic digestion at all WWTPs exceeded 38%, which is one of the major parameters for regulatory compliance with the VAR requirements of the 40 CFR 503 Rule. The results indicate that meeting the 38% VS destruction requirement for VAR compliance is not a useful parameter for predicting odor quality of dewatered biosolids. Further investigation of the relationships between digestion parameters and odor characteristics of the final biosolids product is warranted.

Novak et al. (2003) suggest that two types of biopolymer exist in biosolids, each of which appears to degrade under different conditions, and degradation of activated sludge might be best accomplished by combined aerobic and anaerobic processes. However, the leftover bio-available protein in aerobically digested biosolids may still lead to release of additional labile protein and odor generation. Likewise, anaerobic digestion may not destroy all the protein leading to downstream odors, especially if the biosolids are subjected to various types of dewatering practices.

Residual Biological Activity (RBA): RBA is considered to be a measure of the biological stability of digested biosolids, since RBA represents the potential for further biological activity through endogeneous decay and consumption of the substrate available in the form of volatile acids and protein. Figure 4-34 illustrates the correlation between dewatered biosolids RBA measured at 40 days and odor DT. Figure 4-35 illustrates the same correlation for cake peak sulfur in the sample headspace. Neither plot indicates any apparent correlation between the two parameters. Like VS destruction, RBA is not considered to be a useful tool to predict the odor potential of dewatered biosolids, based on the results of this study.

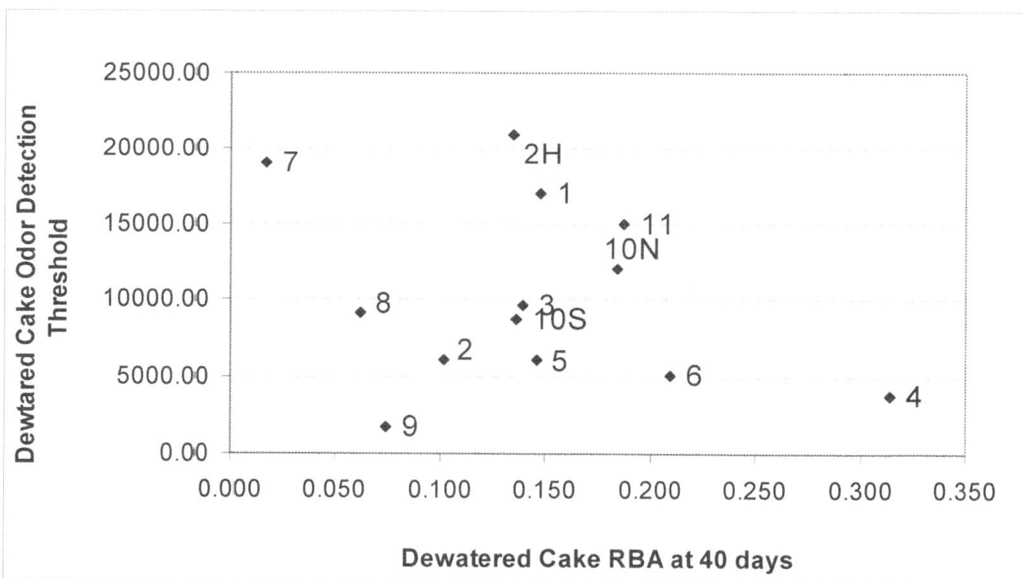

Figure 4-34. Impact of Dewatered Biosolids RBA (40 Days) on Odor Detection Threshold (DT).

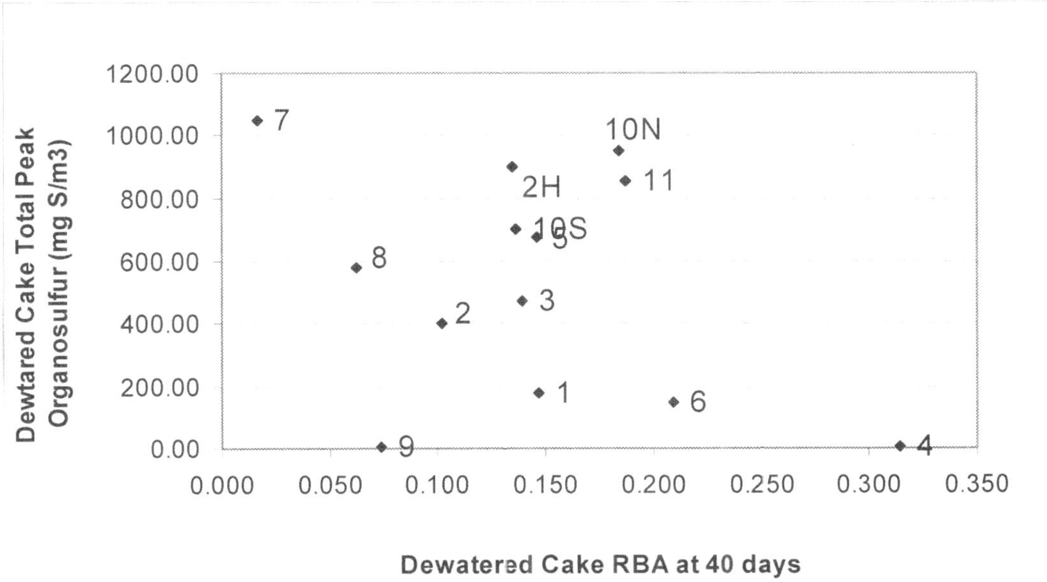

Figure 4-35. Impact of Dewatered Biosolids RBA (40 Days) on Headspace Peak Organosulfur Concentration.

Thermophilic Digestion: Based on the results of a comparison between one test WWTP using thermophilic anaerobic digestion and the other 10 WWTPs employing mesophilic anaerobic digestion, thermophilic digestion produces a different pattern and timing of odor release and different odor constituents, in its dewatered biosolids. The only thermophilic digestion plant in the study, WWTP No. 8, showed significant differences from the other 10 WWTPs in biosolids odor characteristics. Dewatered biosolids from WWTP No. 8 showed the least amount of odor at the seven-day point, but its organosulfur emissions peaked after 30 days of storage. The results from one thermophilic digestion process indicate that thermophilic digestion may remove the precursors of the simpler odor causing compounds, but it also may delay the generation of more complex odorous compounds (such as methanethiol), which can contribute to higher odors after a longer period of anaerobic storage time.

Because the biosolids odor characteristics from WWTP No. 8 were not repeated in the 10 mesophilic WWTPs, it was concluded that thermophilic digestion and its impacts on odor potential and generation patterns warrant further investigation. More research is necessary to better understand odor characteristics of thermophilically digested biosolids. Temperature effects on volatilization of odor compounds also need to be investigated. Conditions in the sample bottles may not have been representative of higher-temperature, onsite storage conditions, which may lead to higher flux of odorous gases. One theory is that delayed peaks could be due to the delayed growth of a new population of microorganisms or re-activation of the existing populations, potentially similar to the ones already established in mesophilic biosolids, responsible from generation of odorous compounds at lower than thermophilic storage temperatures.

Multiple Regression Analysis of Digestion Parameters and Cake Odor: A series of multiple regression analyses were performed on digester related parameters, namely mass of bound protein and iron in digester effluent, digester SRT, fraction of WAS in digester feed, digester VS destruction and digester effluent VFAs, and odor DT and VSCs. No additional correlations were revealed as a result of this analysis (Table 4-8). It is likely that the inability to include the impact of mechanical dewatering contributed to the lack of correlation between the different parameters and odor production. The energy of the dewatering device affects odor, but this energy of dewatering imparted on the biosolids is not quantified and could not be included in the model. Additional research is needed to quantify the impact of mechanical dewatering equipment, especially in cases where high energy forces are acting on the biosolids. This information would provide a possible link to examine the effect of dewatering on odor production and allow better modeling and prediction of odor production during cake storage. It would also provide a better understanding of upstream parameters on odor production.

Table 4-8. Summary of Results from Multiple Linear Regression.

Independent Variables	Dependent Variables	R^2
1. Mass of bound protein 2. Digester Fe (mg/g DS)	DT on Day 6 from cake	0.22
1. Digester SRT (d) 2. Digester Fe (mg/g DS)	DT on Day 6 from cake	0.077
1. Digester SRT (d) 2. Digester Fe (mg/g DS) 3. Fraction of WAS in Digester	DT on Day 6 from cake	0.11
1. Digester Fe (mg/g DS) 2. Digester SRT (mg/g DS)	Peak H_2S from cake	0.17
1. Fraction of WAS in Digester 2. Digester SRT (d)	Mass of bound protein in digester effluent	0.17
1. Fraction of WAS in Digester 2. Digester SRT (d) 3. Digester Fe (mg/g DS)	Mass of bound protein in digester effluent	0.20
1. Digester SRT (d) 2. Digester VS destruction 3. Fraction of WAS in Digester	DT on Day 6 from cake	0.097
1. Digester SRT (d) 2. Digester VS destruction 3. Fraction of WAS in Digester	Peak volatile sulfur compounds	0.38
1. Digester VS destruction 2. Digester effluent VFAs	DT on Day 7 from cake	0.094

4.4.2.3 Conclusions

The project team evaluated the sample analytical results and WWTP operation records to investigate the impact of effective digestion, as defined in terms of low VFAs, sufficiently long SRT, low RBA, and increased VS destruction, on dewatered biosolids odors. The conclusions are:

- Based on the 11 test WWTPs of this study, a well-digested biosolids is not particularly odorous when sampled immediately after digestion. However, downstream handling processes (dewatering, conveyance, etc.) increase the odor potential of biosolids cake.
- VS destruction and RBA may not be useful parameters to measure biosolids odor potential.
- Thermophilic digestion may lead to different biosolids odors development patterns compared to mesophilic digestion, but this is based on data from one WWTP with thermophilic digestion and 10 WWTPs with mesophilic digestion.

4.4.2.4 Recommendations

To better understand the interaction between digestion performance and odor potential in dewatered biosolids, the following additional research needs to be performed:

- Pilot-scale studies where digestion parameters can be varied in conjunction with dewatering parameters to isolate the impact of various digester operation conditions on final product odors, such as the impact of varying digester SRT on biosolids odors, and the impact of digester VS destruction on biosolids odors.
- Processing of biosolids prior to digestion should be investigated for its impacts on biosolids odor potential.
- The thermophilic digestion process should be investigated in relation to achieving a better stabilized biosolids product with acceptable odors.

4.4.3 Impact of Dewatering and Conveyance on Biosolids Odors

4.4.3.1 Hypothesis

Researchers have recently indicated that post-digestion solids dewatering and cake conveyance processes can contribute to odor and VSC release from anaerobically digested biosolids (Erdal et al., 2003; Higgins et al., 2002; Murthy et al., 2002a; Murthy et al., 2002b; Murthy et al., 2003). These researchers suggested that an increase in odor and VSC production occurred subsequent to high-solids centrifugation and certain types of cake conveyance. Other researchers have proposed that biosolids cake pumps also contribute to volatile sulfur production (Ross et al., 2002). One of the potential mechanisms for odor and volatile sulfur production is that solids processing equipment can impart high energy forces on the biosolids cake, resulting in an increase in bio-available protein, a precursor for odor and volatile sulfur production. Another possible mechanism is the inhibition of methanogenesis during dewatering, a biological metabolism necessary for deodorization (Higgins et al., 2003).

4.4.3.2 Results

Survey of MT production from facilities using centrifuges: Figure 4-36 presents the peak MT production from 10 dewatering processes at eight of the 11 WWTPs. Data from WWTPs No. 4 and No. 9 (non-centrifuge dewatering plants) and from WWTP No. 8 (thermophilic digestion plant) were not included in order to minimize the multiple interactions on cake MT production. A linear regression analysis yielded an R^2 value of 0.66 for this data set. The low- and high-solids centrifuges from WWTP No. 2 are separately represented in the figure as 2 and 2H, respectively. MT production for two separate digestion processes for WWTP No. 10 is also shown separately (10N and 10S) in this figure.

WWTPs No. 1, No. 2, No. 3, and No. 6 operate low-solids centrifuges and the MT emissions are correspondingly lower. WWTPs No. 2H, No. 5, and No. 11 operate high-solids centrifuges and the peak MT emissions are much higher. The cakes for the high-solids centrifuge WWTPs on an average contain 5 g more VS in the sample bottles compared to WWTPs with low-solids centrifuges. This represents a 25% increase in VS from the low-solids centrifuge cakes. The corresponding average peak MT for the high-solids centrifuge WWTPs is 187% greater than the low-solids centrifuge WWTPs, indicating a substantially higher VSC production than the increase in VS alone would generate.

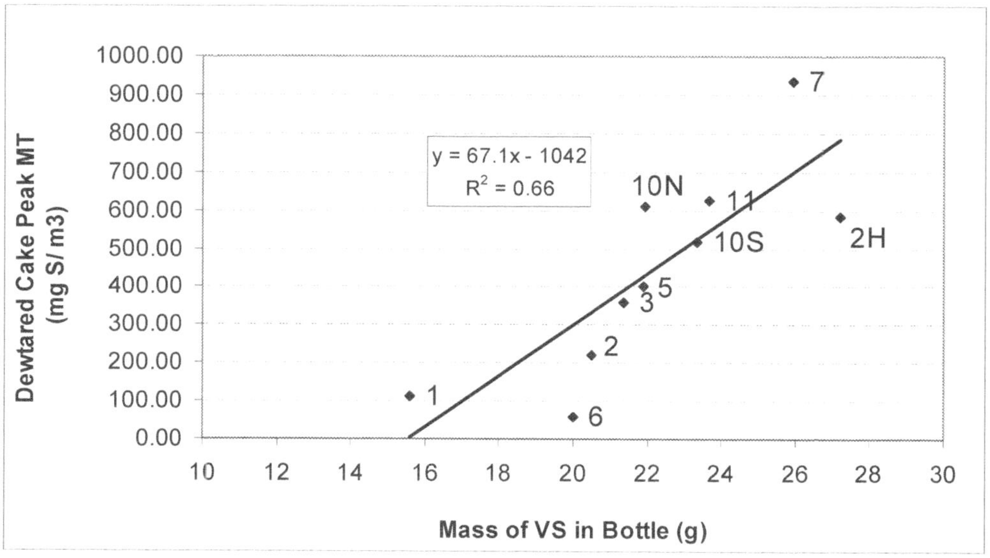

Figure 4-36. Comparison of Peak Headspace MT and Mass of VS in Bottle for 10 Centrifuge Dewatering Processes from Eight WWTPs.

The remaining centrifuges in the study are located at WWTPs No. 10 and No. 7. WWTP No. 10 operates low-solids centrifuges that have been optimized to produce higher-solids cakes. The two dewatered biosolids samples in WWTP No. 10 were 26% and 28% TS, which is closer to the 29% average dry solids for the high-solids centrifuge plants compared to 22% average dry solids for the low-solids centrifuge plants. The cakes from WWTP No. 10 have relatively high peak MT emissions, also residing with the grouping of high-solids centrifuge plants. WWTP No. 7 operates a centrifuge that produces high-solids cake as well (27% dry solids). In addition to the higher cake solids, the biosolids samples in the bottles contained higher VS content than the other WWTPs. The cake from this WWTP produced the highest peak MT emissions. Interestingly, this plant had a relatively low level of RBA.

In summary, a relatively modest increase in VS in the sample bottle (typically due to an increase in dry solids content) between low-solids and high-solids centrifuges resulted in a much more substantial increase in peak MT emissions from the high-solids centrifuges. It follows that increases in VS could result in higher biosolids odors, since MT has been shown to be a significant contributor to biosolids odors.

Side-by-side evaluation of two types of centrifuges at WWTP No. 2: WWTP No. 2 employs new high-solids centrifuges alongside low-solids centrifuges purchased in the mid-1980s. The low-solids centrifuges have been reconfigured with automatic hydraulic back-drives to produce higher cake solids. The dry and volatile cake solids produced by the two centrifuges are listed in Table 4-9. The protein extractions for the two cakes are given in the same table. The difference in percent dry cake solids between the low- and high-solids centrifuges cake samples resulted in different masses of total and VS in the headspace bottles, since the cakes were measured into the bottle on wet mass basis. This increase in mass represents a 30% increase in total and VS in the sample bottle for the high-solids centrifuge cake compared to the low-solids centrifuge cake. The high-solids centrifuge cake contained a higher mass of labile protein in the sample bottle and represented about 80% increase in labile protein in the bottle as compared to the low-solids centrifuge cake. Since the total and VS increased by only 30% between the low-solids and high-solids cake samples, there was an increase in normalized mass of protein in the bottle, expressed as mg/g DS and presented in Table 4-9.

Table 4-9. Solids Characteristics for Low- and High-Solids Centrifuge at WWTP No. 2.

	Dry Cake Solids (%)	Volatile Cake Solids (%)	Labile Cake Protein (mg/bottle)[1]	Labile Cake Protein (mg/g DS)
Low-Solids Centrifuge	27.4	14.3	36	13
High-Solids Centrifuge	35.1	18.7	66	19

[1] Mass of bound protein in sample bottles

Table 4-10 presents the volatile sulfur and odor characteristics for the low- and high-solids centrifuge cakes. An evaluation of the data for the two centrifuges shows over 120% increase in total peak sulfur production for the high-solids centrifuge as compared to the low-solids centrifuge. The difference in odor DT was greater than 240% when analyzed six days after storage. In other words, a 30% increase in cake solids in sample bottles resulted in a much more substantial increase in VSC and biosolids cake odors. Therefore, an increase in dry cake content in the bottle does not appear to sufficiently explain the cause of the increase in volatile sulfur compounds. An increase in labile protein beyond the increase in cake solids content for the high-solids centrifuge cake compared to the low-solids centrifuge cake suggests that this increase in protein contributed to the increase in VSCs.

Table 4-10. Volatile Sulfur and Odor Characteristics for Low- and High-Solids Centrifuges at WWTP No. 2 (All Sulfur Units in mg S/m^3).

	Total Peak Sulfur	Peak H_2S	Peak MT	Peak DMS	Peak DMDS	Peak DMTS	DT Day 6
Low-Solids Centrifuge	405	2.0	220	149	26	8.0	6,100
High-Solids Centrifuge	900	2.0	584	227	74	13	21,000

Impact of conveyance on VSC production: A prior study suggested that VSC production increases with screw conveyance of low-solids centrifuge cake and that VSC production peaks more rapidly for conveyance of high solids centrifuge cake (Murthy et al., 2002b). Similar observations were made by Erdal et al. (2003) for screw conveyors and by Ross et al. (2002) for cake pumping of high-solids centrifuge cakes. In this study, no clear distinctions could be made for samples obtained upstream and downstream of conveyance (see Figure 4-8) to ascertain impacts of this process. In addition, since some of the samples obtained from Sample Location I were obtained from storage silos rather than immediately downstream of the conveyance equipment, the impact of conveyance could not be verified effectively.

4.4.3.3 Discussion

From this investigation, it was difficult to conclusively establish the cause of increase in VSC emissions from biosolids subsequent to centrifugation. Other studies have suggested that it is due to an increase in labile protein. This study found a similar relationship (Subsection 4.3.1), and further suggests that there are two factors influencing the apparent increase in VSC emissions from higher-solids cakes. The first factor may be directly related to an increase in the dry solids content. The second factor appears to be related to the increase in the amount of labile protein present in the bottle (Table 4-9). The mechanism responsible for the increase in the labile protein was not evaluated in this study. However, some prior research indicates that the labile protein can be made bio-available by shear forces applied by different types of equipment (Murthy et al., 2002a; Murthy et al., 2002b; Murthy et al., 2003).

4.4.3.4 Conclusions

- High-solids centrifuges can produce more VSCs than low-solids centrifuges, beyond what is accounted for by the increase in cake TS.
- A variation in torque and differential rpm can lead to dewatered cake characteristics that are different enough to produce different VSC emissions. The opposite is true for a low-solids machine upgraded to produce higher solids, which produce lower VSCs in spite of the solids content in the centrifuge cake.
- No clear distinctions can be made for samples obtained upstream and downstream of cake conveyance devices to ascertain impacts of this process on odor or VSC production.

4.4.3.5 Recommendations

- Further research should be conducted to understand the centrifugation process and its impact as a potential cause for an increase in dewatered biosolids odors.
- Further study is required to evaluate impacts of high-energy conveyance on dewatered biosolids odors.
- Care should be practiced when categorizing centrifuges as low- or high-solids, based on the solids content of the produced cake. The operational parameters, such as backdrive torque and differential rpm, should be further investigated for managing odors in the final product.

4.4.4 Impacts of Biosolids Cake Storage and Time on Odors

4.4.4.1 Hypothesis

Prior research has shown the odor or VSC production from biosolids cake to be time dependant (Murthy et al., 2002 b; Murthy et al., 2003; Higgins et al., 2003) over more than a week of storage. A product that is relatively nonodorous at the WWTP or immediately after dewatering can become odorous during storage. Additionally, research has shown the emission of VSCs to be temperature dependent (Higgins et al., 2003), with an increase in reduced sulfur occurring more rapidly at higher temperatures. The objective of this investigation was to verify some of these time-dependant impacts for the 11 test WWTPs.

4.4.4.2 Results

The stacked bar chart in Figure 4-37 presents an example of VSCs (H_2S, MT, DMS, and DMDS) emitted after anaerobic storage of biosolids from WWTP No. 6, which was found to be representative of most of the test WWTPs. The predominant VSC emissions during storage were MT and DMS. MT production typically peaked prior to DMS peak production. Figures 4-8 and 4-21 illustrate the time-to-peak for VSCs and the overall peaking patterns of reduced sulfur emissions, respectively. As shown in Figure 4-21, the peak VSC production varied considerably from WWTP to WWTP, with maximum VSC emission of greater than 1,800 mg/m^3 and minimum VSC emission of less than 10 mg/m^3. The time-to-peak also varied considerably (Figure 4-8), with a minimum time-to-peak occurring three days after storage, a maximum time-to-peak occurring 35 days after storage, and a median time-to-peak occurring seven days after storage for cakes obtained after dewatering (Sample Location G). Interestingly, the longer 35 days-to-peak VSC production occurred for the thermophilicly digested biosolids cake.

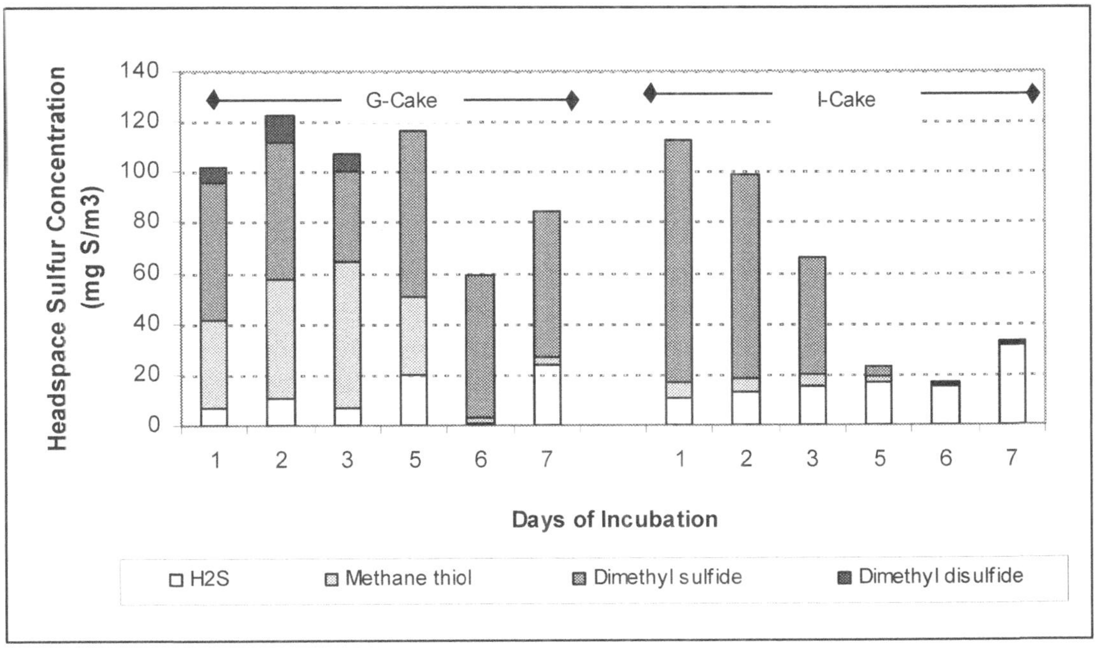

Figure 4-37. Profile of Sulfur Compounds Measured in Post-Dewatering Biosolids Headspace for WWTP No. 6

4.4.4.3 Discussion

Prior research indicated that the formation and emission of MT precedes the formation and emission of DMS (Higgins et al., 2003). It has been hypothesized that the formation of MT occurs mainly from the breakdown of the protein methionine, and the formation of DMS occurs subsequent to MT formation through the methylation of MT. Hence, the formation of DMS lags behind the formation of MT. Furthermore, the methanogens present in digested biosolids are capable of de-methylating these sulfur compounds to inorganic sulfide. Therefore, MT and DMS emissions diminish after peaking. The inorganic sulfide produced after de-methylation was not found in the headspace for most test WWTPs. These ions are thought to have formed in the biosolids cake and precipitated in the presence of metal salts. In summary, a typical odor profile consists of an initial increase in MT, followed by an increase in both MT and DMS, followed by successive depletion of MT and DMS via methylation and demethylation reactions.

4.4.4.4 Conclusions

- Biosolids storage leading to protein and methionine breakdown are significant sources of VSCs emitted from the stored biosolids.
- DMS peaks subsequent to peak MT production from stored biosolids.

4.4.4.5 Recommendations

Additional research is needed for a better understanding of the following topics:

- Determining the cause(s) of the inhibition of methanogenesis in digested dewatered biosolids.

- Determining the impact of VSC deodorization by methanogens for odor mitigation.

4.5 Remarks

In the light of the results, discussion, and conclusions presented in this chapter, it can be concluded in general that odors from biosolids are dependent on a variety of factors. Most of the causative factors appear to be interrelated, potentially leading to emissions of odorous compounds at any point of biosolids treatment or handling, including their transport and ultimate use or disposal.

Digestion of primary and waste activated solids is such a crucial point in the biosolids treatment train that downstream biosolids odors have traditionally been thought to depend primarily on the efficiency of digestion in achieving biosolids stability. Although the results of this study indicate that better digestion may help to reduce biosolids odors, the constituents of the wastewater and the changes it undergoes during treatment, as well as the nature of biosolids handling processes upstream and downstream of digestion, were also found to have significant impacts on biosolids odor quality.

Based on these conclusions, the following practices are recommended:

- Prior to the selection of the biosolids processing equipment, different processes should be evaluated by pilot testing alternative technologies and comparing the odor emissions and odor release patterns in processed biosolids.
- Minimizing available protein can reduce odors.
- VSCs can be used as an odor assessment tool for non-limed biosolids handling.
- Longer digestion times can lead to lower VOSC emissions.
- VS destruction and RBA should not be heavily relied on as indicators of odor potential.
- Since iron, aluminum, calcium, and magnesium appear to bind differently with biosolids and soluble organics, to manage odor precursors their use at different points of liquid and solids treatment train should be tied to downstream biosolids handling practices.

Chapter 5.0 describes the above-recommended practices in more detail and presents an overall summary of the causative effects of various WWTP parameters investigated as part of this study. It provides recommendations for design and operations of biosolids handling systems, as well as guidance for further research. The implications of this study's findings on management practices dealing with biosolids processing and disposal or reuse are also presented in Chapter 5.0.

CHAPTER 5.0

Conclusions and Recommendations

Following the completion of the intensive field sampling and analysis period, the research team compiled and examined the collected data to develop conclusions on the sources of odor generation at wastewater treatment plants. Based on these research findings, which are presented in Chapter 4.0, the conclusions and recommendations of the project research team are summarized in this chapter, along with practical and engineering implications of the findings.

5.1 Hypotheses Developed and Categorized

The hypotheses developed by the project team and listed below are categorized as "conclusive," "inconclusive," or "no correlation," based on the results presented in Chapter 4.0.

5.1.1 Hypotheses Supported Based on Study Results

1. Higher amounts of bio-available protein create more odors in biosolids cake. Protein was measured in three fractions: one soluble and two bound fractions that were extracted from biosolids samples. The "soluble" fraction was measured on filtrate of the liquid biosolids and centrate samples. The two bound fractions, consisting of a mild ("labile fraction") and a more rigorous extraction ("NaOH extract") were measured on the cake samples. The labile fraction is a measure of the bio-available protein, and varied between 14 to 44 mg/g DS in digester effluent samples. The NaOH extract, which varied between 95 to 295 mg/g DS in digester effluent, represents a fraction much harder to serve as a substrate for biological activity. Labile protein was directly correlated (R^2=0.62) with methane thiol, which was the major organosulfur compound that led to in-plant odors at the tested WWTPs.
2. Different dewatering practices affect bio-available protein differently and thereby some practices tend to increase odors in the biosolids cake. For example, the side-by-side comparison of two centrifuge types at WWTP No. 2 revealed that there were enough differences between the two machines to result in a 46% increase in labile protein when the same digested biosolids were dewatered with the high-solids centrifuge compared with the parallel low-solids unit. However, more mechanistic studies are required before definitive relationships can be established between the impacts of various dewatering practices and

increased biosolids odors (high-solids centrifuges that create increased odor from dewatered biosolids when compared to low-solids centrifuges, plate-and-frame presses, or belt-filter presses).

3. VSCs, not H_2S, are the major odor sources in anaerobicly digested biosolids. Concentrations of VSCs were considerably higher than concentrations of H_2S in the headspace samples from nine of the 11 test facilities. In only two plants was the presence of H_2S significant: Dewatered and post-dewatered cake sample VSCs for WWTP No.1 were dominantly (93% and 89%, respectively) H_2S, and at WWTP No. 7 H_2S presence in the bottle head space was 58% of all the VSC. The remaining nine WWTPs had an average of 7% H_2S in all the measured VSC released. A multiple regression analysis showed that the organic VSCs (MT, DMS, DMDS, and DMTS) were very well correlated with odor DT as measured by an odor panel. The correlation coefficient for this relationship is 0.90. These data support findings that show VSCs are responsible for much of the odor, especially in the first week of cake storage. The equation for this relationship is as follows:

$$DT = 3048 + 3.2(MT) - 29.7(DMS) + 105(DMDS) + 1060(DMTS)$$

where:

DT = detection threshold on Day 6;
MT = peak methanethiol concentration (mg S/m^3);
DMS = peak dimethyl sulfide concentration (mg S/m^3);
DMDS = peak dimethyl disulfide concentration (mg S/m^3);
DMTS = peak dimethyl trisulfide concentration (mg S/m^3).

Interestingly, the DMS concentration appears to reduce odors measured by DT as shown by the negative term on the coefficient. This contradicts some previous research that showed DMS as an important odor source, although the complex interactions of the different species in the model might have contributed to these results.

4. For most mesophilicly digested biosolids cake, odor concentrations rise and then decline over time during storage. The reduced-sulfur compounds, which are the primary constituents of biosolids odors, peak in a somewhat predictable manner. Methanogenesis, which consumes organic sulfur compounds, potentially contributes to the somewhat predictable rise and decline of biosolids odors during storage. Although they varied from plant to plant, the VSCs measured in all the mesophilic samples peaked substantially earlier than the peaks observed in the thermophilic samples. Mesophilicly digested dewatered cake samples peaked within one to 14 days of storage, whereas the thermophilic sample peaked on the 35th day of storage.

5. Thermophilicly digested biosolids have a different pattern of time release of odor constituents than mesophilicly digested biosolids. The time accumulation and release pattern of odors from thermophicly digested biosolids cake appears to be considerably longer (35 days) than odors from mesophilicly digested biosolids, with a smaller proportion of its odor strength due to VSCs. The peaks observed on samples obtained from WWTPs with mesophilic digestion were higher (2,408 mg S/m^3 versus 621 mg S/m^3). The study included only one WWTP with thermophilic digestion (even though its biosolids odor characteristics were different than the other 10 mesophilic WWTPs). For this reason this conclusion needs to be confirmed through other studies that incorporate WWTPs with thermophilic digestion, and further studies on thermophilic digestion and its relation to odors.

6. The study shows that iron in sufficient concentrations binds bio-available protein in biosolids cake, and thus reduces odor production of dewatered biosolids. The presence of iron in digested biosolids correlated with lower concentrations of bio-available protein in most digested biosolids samples, and there was evidence of odor reduction in most dewatered cake samples that had significant concentrations of iron. However, the extent of the influence of iron and other cations on dewatered biosolids odors needs to be evaluated through controlled, mechanistic studies.

5.1.2 Hypotheses Rejected Based on Study Results

The list below represents hypotheses that were developed because the team believed that potential relationships might exist, but data collected indicated that no correlations were present based.

1. The study found no relationship between influent sulfate concentrations and odor production in biosolids cake. Any relationship that may exist between sulfate in the wastewater and odors from dewatered biosolids is overcome by other components of the wastewater or other influences during wastewater treatment and biosolids handling.
2. The study findings provided no evidence that WAS has a higher odor potential than primary sludge following digestion. To the contrary, it appeared based on one correlation (No. 5 above) that higher percentages of WAS in digester feed could possibly lead to lower odors in digested biosolids cake.
3. The study findings provided no evidence that enzyme activity can be used as an indicator of biosolids odor production. The data collected did not show any correlation between the enzyme activity and VSC emissions from the biosolids cake.

5.1.3 Hypotheses Found to be Inconclusive Based on Study Results

The list below represents hypotheses that were developed because the team believed that potential relationships exist. While some results of the study supported these relationships, a conclusive relationship could not be established based on all of the data. Further study is recommended to either confirm or refute these otherwise inconclusive relationships.

1. The study results did not show that high-energy (or pressure) conveyance of biosolids after dewatering (through screw conveyors or cake pumps as opposed to belt conveyors) leads to increased odors in the biosolids cake
2. The study results did not show that cation concentration of the influent wastewater affects the odor production of dewatered biosolids. A general positive trend was found (Figure 4-17) when the monovalent to multivalent cation ratio (M/D) was plotted against digester soluble protein (mg/L), but for the M/D values exceeding 0.5, the variability increased significantly. For this reason it was decided that there was not enough evidence to prove or disprove this hypothesis.
3. The study results did not show that mixing primary sludge and WAS prior to digestion has any definitive impact on odors in digested, dewatered biosolids. The percentage of WAS in digester feed varied between 31-65% for the WWTPs evaluated during this study. Two of the WWTPs (No. 6 and No. 7) employ combined thickening of primary and secondary biosolids. WWTP No. 6 did not have significant emissions of odor (255,000 DT) or VSCs (139 mg S/m^3), but WWTP No. 7 exhibited one of the highest cake odors (950,000 DT) and VSC

emissions (2,408 mg S/m^3) in the study. The project team found three major process differences between these two WWTPs: primary sludge detention time, digester SRT, and the ratio of WAS to primary sludge in the digester feed. Digesters operating at 28 days SRT at WWTP No. 6 were receiving 54% WAS, whereas the digesters at WWTP No. 7 were operating at 14 days SRT and receiving 33% WAS. Primary sludge age was long (83 hours) for WWTP No. 6 compared to the one-hour primary sludge detention time of WWTP No. 7. One correlation showed organosulfur emissions decreasing with increasing WAS content of the digester feed at WWTPs with centrifuge dewatering, which contradicts the hypothesis that higher amounts of WAS in the digester feed lead to more odorous biosolids. This hypothesis therefore requires further study before any conclusions can be drawn.

4. The study results did not show that metal cations such as iron in the biosolids can reduce odors in the biosolids cake, post-dewatering. The total iron concentration measured in the digesters of the 11 WWTPs varied between 233 mg Fe/L and 2,200 mg Fe/L. These concentrations represent all the iron present in the sample, including iron already bound in complexes with sulfur, phosphorus, and organic material, and free iron available for further complexing. A correlation was observed between increasing amounts of digester total iron and decreasing amounts of bio-available protein in digested biosolids, but the correlation did not extend to decreased odors in the biosolids cake. The lack of knowledge about the bioavailability of proteins and other organic material complexed with cations such as iron and aluminum caused the project team to remain undecided on this hypothesis. Also, the fate of the protein that forms complexes with metal cations needs to be investigated, since de-complexing would lead to downstream odor production.

5. The study results did not show that longer primary sludge detention time prior to digestion created higher odors in digested biosolids. In conjunction with Item No. 4 above, the data were not conclusive. The WWTP RFI data showed a wide range of primary sludge detention times (one to 214 hours), but no correlation could be developed between primary sludge detention time and odor levels in the biosolids cake.

6. The study results did not show that longer WAS SRT in the treatment process produces less odors in digested dewatered biosolids. Most of the SRT values were in the range of one to 10 days for the WWTPs examined, with 75% of the WWTPs operating at WAS SRTs below five days. WWTP No. 5 (high-solids centrifuge) and WWTP No. 7 (low-solids centrifuge), operating at a WAS SRT of equal or greater than 4.5 days, for example, exhibited organosulfur emissions of 678 mg S/m^3 and 1,047 mg S/m^3, respectively. It is evident that WAS SRT needs to be further evaluated in regard to its potential impact on odor generation.

7. The study results did not show that anaerobic digestion achieving 38% minimum volatile solids destruction (thus meeting the Part 503 Rule's requirements for VAR under Option No. 1) reduces odors in dewatered biosolids. With respect to the change in protein (mg/g DS) across the digestion process, it was observed that the labile- and NaOH-extracted proteins appeared to be degraded as expected. The net change in labile protein in digesters at mesophilic plants varied between 0 and 71%, and the thermophilic plant had a negative net change in labile protein across the digester. This and other negative values may be indicative of phase changes between the two fractions of protein.

All digesters evaluated in this study appeared to be in full compliance with VAR standards, based both on VAR Option No. 1, which requires 38% VS reduction in digestion, and VAR Option No. 2, which is an indicator of low RBA based on the AVSR test as described under that option. In addition, all digested biosolids had relatively low odors in the liquid phase. Odors from biosolids increased significantly after dewatering in this study, however, especially if the dewatering was performed by centrifuges.

8. The study results did not show that TMAs are significant sources of odor in digested/non-limed biosolids. The data set collected during this study did not show any significant TMA emissions. The total nitrogen emissions were only between 0 mg N/L and 4.34 mg N/L for post-digestion sampling points. Since most nitrogen-bearing compounds are not as volatile as the VSCs and harder to measure, however, the team decided that insufficient data on TMA was collected to substantiate or refute this hypothesis.

5.2 Bottle Headspace Sampling and Biosolids Odor Analysis

One conclusion from the foregoing discussion of the results is that the headspace method used as the method of biosolids storage, incubation, and odor analysis can be reliably applied to other studies, since it was found to successfully simulate actual biosolids storage conditions. The headspace sampling and analysis method employed in this study allows for reasonable comparison of potential odor emissions from biosolids. The purpose in using this method was to eliminate the impacts of changing conditions of emission flux, atmospheric dilution, and oxygen on odorous compound generation potential. These factors impose rapid changes on emissions, making it difficult to trace released compounds back to their sources in wastewater and process parameters.

The bottle headspace method was investigated during the first phase of this study and during the "dry run" sampling event, and it was selected over air sampling, flux chamber, and the purge-and-trap sampling methods. As explained in Section 4.2.1, the headspace method eliminated the dilution of the odors and facilitated a full picture of odorous compound emission potential from the anaerobic core of biosolids piles to be drawn. The olfactometry measurements were found to yield very high results (in the range of 4,750 to 1,050,000 DT and 3,250 to 700,000 RT). The headspace samples used for these measurements were a result of a six-day-long accumulation of all the released gaseous compounds, instead of resulting from grab sampling that is more representative of an instant odor "snapshot" rather than the odor potential of the biosolids samples.

Considering the anaerobic core of biosolids storage piles, a sealed bottle that does not allow air penetration to the cake sample, yet captures the released gases, is a valuable tool for similar applications. It also allows absorption and re-assimilation of the released odorous chemicals by the active microbial population present in the cake samples. The sample bottles serve as containers for sample shipment, preserve the cake in its anaerobic state (as in a cake pile or biosolids storage hopper), serve as bench-scale cake incubators to simulate full-scale cake storage, and provide containment of headspace gases that characterize the odor of the biosolids in equilibrium with the biosolids sample for odor analysis. They also prevent any further handling of the samples.

Another facet of the study focused on the relationship between the olfactometry measurements performed on the Day 6 headspace samples and VSC emissions. Odors correlated with the concentration of VSCs analyzed on the same day, for most of the WWTPs in the study. For a few of the WWTPs, the correlation between odors and VSC emissions was not as clear, which could be attributed to the relationship between odors and nitrogen that was not sufficiently evaluated during this study due to low organic nitrogen in the sample headspaces.

5.3 Engineering Implications of Study Results

A major objective of this study was to identify the potential sources and mechanisms resulting in odor generation from anaerobicly digested and dewatered biosolids. In theory, once odor sources and mechanisms are identified, it should be possible to design improvements to plant processes or implement operational changes that can reduce the odor impacts from biosolids. This section discusses the following two general areas of the study results that have engineering implications:

1. Constituents of the wastewater: This category is considered an engineering implication because constituents of wastewater or biosolids can be changed during the treatment process by chemical addition or through process enhancements.
2. WWTP design and operating parameters: By their nature, design and operating parameters are determined by human decisions made in the design process or during daily WWTP operations.

5.3.1 Influence of Wastewater and Biosolids Constituents on Biosolids Odors

Bio-available protein in the volatile solids fraction of the cake solids was found to be the main source of the odorous compounds in anaerobicly digested and dewatered biosolids. It was hypothesized that amino acids, as building blocks of proteins and the bearers of the sulfur and nitrogen in the microbial cells, can lead to formation of odorous compounds when degraded during endogenous breakdown of volatile solids. Cysteine and methionine are sulfur-containing amino acids that have been shown contribute to biosolids VSC emissions.

A correlation was found (R^2=0.62) between the labile protein fraction of the biosolids and MT emissions. An even stronger correlation between cake methionine and headspace MT concentration (R^2=0.96) was found, confirming the validity of the hypothesis that bio-available protein leads to biosolids odors. A relationship was also found between cake methionine and odor DT. This finding should lead to further study on the potential for decreasing odors in anaerobicly digested biosolids by degrading proteins prior to or during digestion.

It was also hypothesized that the activity of the protein degradation enzymes (peptidases) should correlate with odor generation, but no relation between the selected peptidase (l-leucine aminopeptidase, a universal proteolytic enzyme activity) and biosolids odors could be confirmed in this study. It was concluded that other factors influencing odor generation have stronger impacts in shaping biosolids odor release compared to the enzyme activity levels measured in the biosolids samples. Based on the data collected, all the samples showed positive activity for the same proteolytic enzyme, with values varying between 0.080 and 1.043 units/mg labile protein for digester samples, and between 0.290 and 1.148 units/mg labile protein for cake samples. It was therefore concluded that enzyme activity is an impractical tool to predict or measure for biosolids odor generation.

Cations are important constituents of wastewater and have significant impacts on floc stability of biological solids, because multivalent cations impart ionic bridging between floc particles while monovalent cations break up these bridges, leading to de-flocculation. Based on the relationship between cations and exocellular polymers, it was hypothesized that biopolymers tied up by ionic bridges should be less available for degradation, thereby reducing biosolids odor formation. Iron as a cation poses a special condition in that it complexes with sulfur-containing compounds, preventing their volatilization.

The investigation did not reveal any correlation between cations and odor generation from the cake samples, but iron in digested biosolids correlated relatively well with cake protein. The correlation between digester cation concentrations and available protein in digested biosolids implies that adjusting the cation balance or iron feed in digesters can decrease protein availability. However, binding of protein with cations also has the potential to concentrate protein in the processed biosolids, which can lead to odors in the biosolids cake after a period of time, if the protein in complex with the metals become bio-available again.

Another hypothesis investigated as part of this study was whether high plant influent sulfate concentration lead to increased odor production, especially in the form of H_2S. This study could not find conclusive evidence to substantiate this hypothesis.

5.3.2 Influence of WWTP Design and Operation on Biosolids Odors

As part of the study, a large data set of operating parameters (RFI data) was contributed by the staff of participating WWTPs and reviewed by the project team. The RFI data contains detailed process information with emphasis on biosolids handling processes. In the search for relationships between process parameters and biosolids odors, mixing of primary sludge and WAS prior to digestion was hypothesized to increase the biosolids cake odors, with the proportion of WAS potentially having the larger impact. Other hypotheses considered the impact of primary sludge detention time and WAS SRT on biosolids odors.

Although the sampling points were chosen to represent similar locations in the WWTP process train, a number of the WWTPs had different sample points in pre-digestion processes. After the differences in sample location were taken into account, the data indicated that mixing primary sludge and WAS can increase odors from liquid biosolids before and after digestion, but the data do not show that WAS has a higher potential for odor generation. Neither detention time for primary sludge or WAS SRT appeared to have any significant relationship with biosolids odors. Therefore in-plant operational parameters should be further examined to investigate the potential impacts of digestion on downstream biosolids odors.

Efficient anaerobic digestion and stabilization of biosolids in accordance with the 503 Rule is generally believe to lead to reduction of odors in digested biosolids. Odor data collected from sampling points prior to and following anaerobic digesters showed that odors emitted from digester effluent for all plants was relatively minimal compared to odors from the same biosolids after dewatering. The results of this study show that conventional biosolids stability criteria are not sufficient to reduce odors in digested, dewatered biosolids. Attempts to correlate digester volatile solids destruction or residual biological activity with biosolids odors or VSC emissions did not reveal any conclusive results. While the 38% VS destruction requirement was exceeded at all test facilities (VS reduction varied between 42-67%), the digested dewatered biosolids still produced high odor and VSC emissions. Olfactometry performed on headspace gas samples of digested biosolids stored for six days showed values as high as 1,050,000 DT, whereas the highest peak total Sulfur was 2,408 mg S/m^3.

Digester SRT, which varied between 14 and 40 days for the 11 plants (75% of the plants had digester SRT values greater than 20 days) was another parameter investigated, and when all the plants were considered, no correlation was found between digester SRT and dewatered cake odor DT. When the plants with mesophilic digestion followed by centrifuge dewatering were considered, a correlation (R^2=0.62) was found between the digester SRT and cake total peak organosulfur concentration. Post-digestion processing and handling of anaerobicly digested biosolids further increased odor generation in dewatered biosolids. A side-by-side comparison of low-solids and high-solids centrifuges at one of the WWTPs showed a greater impact on release of the bio-available protein (46% more labile protein) from dewatered biosolids, resulting in a 125% greater VSC release from the biosolids dewatered by the high-solids centrifuge. Olfactometry measurements also showed 244% higher DT values in the high-solids centrifuge cakes from this WWTP.

One of the plants uses anaerobic lagoons to store digested biosolids for approximately six months in a typical year. Samples collected from those lagoons released the lowest odors (3,500 DT) and VSC concentrations (60 mg S/m^3) in the sample bottle headspace of any of the WWTPs investigated. Storage of biosolids cake did appear to further increase odors at WWTP Nos. 1, 5, 8, 9, and 10 when odor emissions from cakes at sample locations G and I were compared. When VSC concentrations were compared, only WWTPs No. 9 and No.10 showed increased VSC increase post storage. The findings of this study therefore indicate that biosolids storage after digestion can affect its odor potential but not as considerably as the impacts of upstream biosolids treatment and handling processes at the WWTP.

There are two main conclusions that can be drawn from the foregoing discussion:

1. There is some evidence, based on the correlation of bio-available protein with biosolids VSC emissions, that digestion improvements leading to more complete destruction of complex organics including proteins can help reduce odors from the biosolids after dewatering.
2. There is further evidence that every biosolids handling process can have an impact on odors in the final biosolids product. Successful anaerobic digestion is not sufficient to maintain low biosolids odors if upstream or downstream processes are imposing conditions which increase the potential for odor generation.

The study included one WWTP with thermophilic anaerobic digestion of biosolids. Although the dewatered biosolids from this WWTP indicated decreased odor generation within the seven-day time period of maximum odor release observed at mesophilic plants, the maximum VSC release from thermophilicly digested and dewatered biosolids occurred at a much later time (35 days of storage) at comparable concentrations (621 mg S/m^3, where the average for the centrifuge cake total peak sulfur values was 906 mg S/m^3). For the nitrogenous odor constituents, this period-to-peak may even be longer for thermophilic biosolids. Based on these results it was concluded that odors from thermophilicly digested biosolids can be generated after the biosolids are taken to the final disposal or application site. These findings regarding thermophilicly digested biosolids should be further investigated, and better means of defining the stability of dewatered biosolids need to be developed for minimizing biosolids odors.

5.4 Recommendations for Additional Research

Based on the conclusions reached and hypotheses not proven as a result of this study, the project team prepared the following list to identify research needs to better understand the generation of biosolids odors:

1. Effects of cations and anions in plant influent need to be investigated using controlled experiments.
2. Controlled investigation of digestion enhancements through mechanical or thermal pre-treatment of digester feed is needed, focusing on destroying enough releasable proteins to reduce odors. Keeping proteins bound through the process is another alternative.
3. Further investigations are needed to optimize pre-digestion processes to minimize odor potential in biosolids cake. Minimization of biosolids odors for high-energy biosolids handling equipment (high-solids machines) need to be investigated, considering that many WWTPs use such equipment due to its superior dewatering capabilities. The correlation between the energy imparted on biosolids during dewatering and biosolids odors, and the threshold of energy and mechanism that enhance odors, needs to be studied to optimize the dewatering processes in conjunction with the digester enhancements.
4. The wastewater industry needs to confirm the validity of headspace odor and chemical analysis to biosolids that are land applied and to extend the capabilities of the headspace method to measure a wider range of N-bearing compounds. Since the nitrogenous compounds are not as volatile, they have the potential to linger on the biosolids surface even after the sulfur compounds are volatilized or consumed. For this reason, N-bearing compounds should also be monitored, which also heightens the need for developing a further improved measurement method.
5. Further research is needed to investigate biosolids odors at specific WWTPs (with mechanistic studies testing these hypotheses) using some of the full-scale data from this study.

Using the strategy of destroying the maximum possible amount of releasable organics and proteins, the research team has decided that the digestion processes can be the pivot point in odor abatement if they can be taken further along the VS destruction curve. For this purpose, a variety of pre-digestion processes for more complete odor precursor destruction is recommended to be investigated for the enhancement of digestion efficiency. In this context, mesophilic and thermophilic digestion need to be further investigated through mechanistic studies where the operation conditions can be varied. Also, appropriate processes need to be paired to investigate biosolids processing and handling combinations downstream of digestion, which should result in a better stabilized and less odorous biosolids product.

5.4.1 Proposed Laboratory-Scale Digestion Enhancement Studies

Laboratory studies with bench-scale simulations of the processes chosen for digestion enhancement are envisioned as follows:

1. *Develop complete list of candidate processes and mechanisms for digestion enhancement:* This initial step would consist of a brief review of recent (past two years) technical literature to identify processes and mechanisms that show promise for enhancement of anaerobic digestion to reduce odors. These processes and mechanisms are generally categorized as

chemical, mechanical, or thermal processes that enhance cell lysis to increase the capability of digestion to degrade the precursors of the odorous compounds. Examples include but are not limited to:

- Pre-pasteurization
- Pre-digestion chemical treatment
- Thermal pre-processing
- Pre-digestion mechanical cell-lysing (sonication)
- Two-phase digestion (several possible types)
- Thermophilic anaerobic digestion (several possible types)
- Anaerobic digestion followed by aerobic digestion (mesophilic aeration)

Most, if not all, of the above-mentioned processes can be simulated on a bench-scale in the laboratory.

2. *Refine candidate list to several processes that can be tested on a bench-scale basis:* Based on the information gathered during the first step, processes can be screened to select the most cost-effective and feasible processes for further study in the laboratory.
3. *Perform bench-scale process work and analytical tests, and document results according to the test protocols.*

5.4.2 Proposed Full-Scale or Pilot-Scale Digestion Enhancement Studies

Based on the work completed as part of the laboratory studies, the steps of pilot- or full-scale studies are envisioned as follows:

1. *Develop site-specific protocols for selected full-scale or pilot-scale tests.*
2. *Implement and stabilize the chosen demonstration processes at the TCR sites*: Equipment should be set up as needed to run the selected processes onsite at either full, partial, or pilot scale. The process must be representative of full-scale installation at that site and must be integrated into the WWTP's other treatment processes.
3. *Perform sampling and analytical work according to the test protocol:* The full-scale analytical protocol should simulate the laboratory-scale protocol as much as possible, so that results can be compared and the effects of scale-up can be evaluated.
4. *Analyze the full-scale data, evaluate, and compile the results of laboratory- and full-scale studies in a report:* Data from full-scale testing need to be statistically analyzed to establish relationships between the tested processes biosolids odors released during normal storage periods.

APPENDIX B: SAMPLE CUSTODY PROTOCOL

Sample possession during all testing efforts must be traceable from the time of collection until the results are verified and reported. Sample custody procedures provide a mechanism for documentation of all information related to sample collection and handling to achieve this objective.

The WERF Study Team leader at the site will be responsible for seeing that the field team adheres to proper custody and documentation procedures for all sampling operations. Chain-of-Custody forms will be used as the primary documentation mechanism to ensure that information pertaining to samples is properly recorded. Copies of the Chain-of-Custody forms and the field logs will be retained in the project file.

B.1 Documentation Procedures

B.1.1 Field Records

Field personnel will be required to keep accurate written records of their daily activities in a bound logbook. All entries will be legible, written in waterproof ink, and contain accurate and inclusive documentation of an individual's field activities, including field data and observations, any problems encountered, and actions taken to solve the problem. The type of data recorded in the field logbook includes field measurements, ambient conditions, and any other information pertinent to sample collection.

Entry errors or changes will be crossed out with a single line, dated, and initialed by the person making the correction. Entries made by individuals other than the person to whom the logbook was assigned will be dated and signed by the individual making the entry. Field logbooks will be available for review by interested parties.

B.1.2 Sample Labeling

Each sample collected will receive a sample label that identifies the sample by a unique sample identification number. These labels are affixed to the sample container prior to sample collection. The sample label shall be sealed on the bottle with clear plastic tape. The sample labels will contain the following information:

Date sample was taken
Sample site
Sample Location ID
Analyte(s)
Sample Number

Examples of preprinted sample labels are provided in Figure B-2.

B.1.3 Sample Master Logbook

A sample master log will be maintained for all samples collected. Each sample will be assigned a unique identification number; a full description of the sample, its origin, and disposition will be included in the log entry.

B.2 Chain-of-Custody Procedures

After the samples are collected and documented in the master logbook, a Chain-of-Custody form will be completed and will accompany the samples to the laboratory (a sample form is provided in Figure B-1). Team members collecting the samples are responsible for the care and custody of the samples until they are transferred or dispatched to the appropriate laboratory. When transferring samples, the individuals relinquishing and receiving the samples will sign, date, and note the time on the record. This record documents sample possession from collection to the laboratory sample control center.

When the samples are received by the laboratory, the sample control officer will verify the Chain-of-Custody form against the samples received. If any discrepancies are observed, they will be recorded on the Chain-of-Custody form and the filed team leader will be notified to correct the problem.

B.2.1 Shipment

All sample shipments will be accompanied by the Chain-of-Custody record, which identifies the contents of each crate. The person relinquishing the samples to the laboratory will request the signature of a laboratory representative to acknowledge receipt of the samples Sample collection and shipment will be coordinated to ensure that the receiving laboratory has staff available to process the samples according to method specifications. All shipping containers will be secured for safe transportation to the laboratory. The method of shipment, courier name(s), and other pertinent information is entered in the "Remarks" section when the samples are to be shipped (i.e.. Federal Express, Express Mail, etc.).

B.2.2 Sample Handling Procedures

The objective of sample handling procedures is to ensure that samples arrive at the laboratory intact, at the proper temperature, and free of external contamination. Liquid and bag samples will be shipped via Federal Express to the appropriate laboratory by field sampling personnel. Each sample shipping container that contains samples for headspace analysis will have an enclosed temperature data logging device in it.

Once the samples have been collected, the methods specify preservation, storage requirements and holding time limitations. Table B-1 summarizes the types of sampling containers to be used and the preservation requirements for the types of analysis to be performed using this General Testing Protocol. It is the responsibility of the sample preparation team and the laboratories to provide these requirements.

Table B-1. Parameters for Sample Preservation WERF Project 00-HHE-5 Phase II Odor Study.

Analysis	Storage Container	Maximum Holding Time and Temp (°C)
Headspace Analysis	500-mL PETE Bottle (141 gram sample)	1 day at 20 ± 5°C followed by 6 days at 22°C storage
Olfactometry	500-mL PETE Bottle (141 gram sample)	1 day at 20 ± 5°C followed by 3 days at 22°C storage
Organics, Cations, and Anions	Nalgene® Bottles @ Sizes Spec'd in Table 5-1 (80% full)	4 days at 0 ± 2°C
RBA	500-mL PETE Bottle (35 gram sample)	1 day at 0 ± 2°C followed by 40 days at 35°C storage

B.3 Calibration Procedures and Frequency

Information is presented in this section pertaining to the calibration of offsite and onsite field instrumentation. Included are descriptions of each procedure or references to applicable standard operating procedures (SOPs), the frequency of calibrations, and the calibration standards to be used.

B.3.1 Laboratory Instrument Calibration

Laboratory instruments are calibrated according to method specifications and are in compliance with the analytical method requirements. Detailed calibration procedures and recommended frequencies are included along with the analytical SOPs, which can be found in Appendix C of this General Testing Protocol.

B.3.2 Field Instrument Calibration Procedures

An important function in maintaining data quality is the check-out and calibration of all field instrumentation. Using referenced procedures, the instruments will be calibrated prior to field sampling. These results will be properly documented and retained. A discussion of the procedures used to calibrate equipment is presented below, and calibration frequency is presented in Table B-2.

Table B-2. Calibration Frequency of Field Sampling Equipment WERF Project 00-HHE-5 Phase II Odor Study.

Sampling Equipment	Calibration	Frequency
Jerome 631-X	Factory	Annual
	Regenerate	Daily
	Zero	Daily

Environmental Management Consulting
Chain of Custody Record

Odor

Form Serial Number **EMC F1-01301**	For Information Regarding These Samples Please Contact: **Thomas R. Card** Environmental Management Consulting 100 292nd Avenue SE, Fall City, WA 98024 425-222-6531 Fax: 425-222-6527 E-Mail: tom_card@msn.com	Client Address and Phone Number ***10844 Ellis Avenue*** ***Foutnain Valley, CA 92708 714-593-7084***	Laboratory Name ***St. Croix Sensory***
Client Name ***OCSD***		Analysis Requested	Laboratory Address ***3549 Lake Elmo Drive*** ***Lake Elmo, MN 55402***
Project Manager ***Chloe Dao***			Laboratory Phone ***651-439-0177***
Requested Completion Date			Laboratory Contact ***Deb Mathies***

Station Number	Date	Time	COMP	GRAB	Sample ID Number	# OF CONTAINERS	Sample Container: Bag	Can	Tube	***ASTM E679-91***										Remarks
	05/22/2000			x	F-101	1	X			X										
	05/22/2000			x	F-102	1	X			X										
	05/22/2000			x	F-103	1	X			X										
	05/22/2000			x	F-104	1	X			X										
	05/22/2000			x	F-105	1	X			X										
	05/22/2000			x	F-106	1	X			X										
	05/22/2000			x	F-107	1	X			X										
	05/22/2000			x	F-108	1	X			X										
	05/22/2000			x	F-109	1	X			X										
	05/22/2000			x	F-110	1	X			X										
	05/22/2000			x	F-111	1	X			X										
	05/22/2000			x	F-112	1	X			X										
	05/22/2000			x	F-113	1	X			X										
	05/22/2000			x	F-114	1	X			X										
	05/22/2000			x	F-115	1	X			X										
	05/22/2000			x	F-116	1	X			X										

Sampler	Date/Time	Relinquished by	Date/Time	HAZWRAP/NEESA Y N QC Level 1 2 3	
Recieved by	Date/Time	Relinquished by	Date/Time	COC	Ice
				Ana Req	Temp ***AMB***
Recieved by	Date/Time	Relinquished by	Date/Time	Cust Seal	pH
				Sample Condition	
Recieved by Laboratory	Date/Time	Sample Shipped Via UPS FEDEX DUO Hand Other	Air Bill No.		
Remarks					

Figure B-1. Example Chain of Custody Form.

WERF Project#00-HHE-5 Biosolids Odor Quality Test Site 1 June 1, 2002 Location **Influent** **Headspace** Sample 1	WERF Project#00-HHE-5 Biosolids Odor Quality Test Site 1 June 1, 2002 Location **Influent** **Headspace** Sample 2	WERF Project#00-HHE-5 Biosolids Odor Quality Test Site 1 June 1, 2002 Location **Influent** **Headspace** Sample 3
WERF Project#00-HHE-5 Biosolids Odor Quality Test Site 1 June 1, 2002 Location **Influent** **Headspace** Sample 4	WERF Project#00-HHE-5 Biosolids Odor Quality Test Site 1 June 1, 2002 Location **Influent** **Headspace** Sample 5	WERF Project#00-HHE-5 Biosolids Odor Quality Test Site 1 June 1, 2002 Location **Influent** **Headspace** Sample 6
WERF Project#00-HHE-5 Biosolids Odor Quality Test Site 1 June 1, 2002 Location **A** **Headspace** Sample 1	WERF Project#00-HHE-5 Biosolids Odor Quality Test Site 1 June 1, 2002 Location **A** **Headspace** Sample 2	WERF Project#00-HHE-5 Biosolids Odor Quality Test Site 1 June 1, 2002 Location **A** **Headspace** Sample 3
WERF Project#00-HHE-5 Biosolids Odor Quality Test Site 1 June 1, 2002 Location **A** **Headspace** Sample 4	WERF Project#00-HHE-5 Biosolids Odor Quality Test Site 1 June 1, 2002 Location **A** **Headspace** Sample 5	WERF Project#00-HHE-5 Biosolids Odor Quality Test Site 1 June 1, 2002 Location **A** **Headspace** Sample 6
WERF Project#00-HHE-5 Biosolids Odor Quality Test Site 1 June 1, 2002 Location **B** **Headspace** Sample 1	WERF Project#00-HHE-5 Biosolids Odor Quality Test Site 1 June 1, 2002 Location **B** **Headspace** Sample 2	WERF Project#00-HHE-5 Biosolids Odor Quality Test Site 1 June 1, 2002 Location **B** **Headspace** Sample 3
WERF Project#00-HHE-5 Biosolids Odor Quality Test Site 1 June 1, 2002 Location **B** **Headspace** Sample 4	WERF Project#00-HHE-5 Biosolids Odor Quality Test Site 1 June 1, 2002 Location **B** **Headspace** Sample 5	WERF Project#00-HHE-5 Biosolids Odor Quality Test Site 1 June 1, 2002 Location **B** **Headspace** Sample 6
WERF Project#00-HHE-5 Biosolids Odor Quality Test Site 1 June 1, 2002 Location **C** **Headspace** Sample 1	WERF Project#00-HHE-5 Biosolids Odor Quality Test Site 1 June 1, 2002 Location **C** **Headspace** Sample 2	WERF Project#00-HHE-5 Biosolids Odor Quality Test Site 1 June 1, 2002 Location **C** **Headspace** Sample 3
WERF Project#00-HHE-5 Biosolids Odor Quality Test Site 1 June 1, 2002 Location **C** **Headspace** Sample 4	WERF Project#00-HHE-5 Biosolids Odor Quality Test Site 1 June 1, 2002 Location **C** **Headspace** Sample 5	WERF Project#00-HHE-5 Biosolids Odor Quality Test Site 1 June 1, 2002 Location **C** **Headspace** Sample 6
WERF Project#00-HHE-5 Biosolids Odor Quality Test Site 1 June 1, 2002 Location **D** **Headspace** Sample 1	WERF Project#00-HHE-5 Biosolids Odor Quality Test Site 1 June 1, 2002 Location **D** **Headspace** Sample 2	WERF Project#00-HHE-5 Biosolids Odor Quality Test Site 1 June 1, 2002 Location **D** **Headspace** Sample 3
WERF Project#00-HHE-5 Biosolids Odor Quality Test Site 1 June 1, 2002 Location **D** **Headspace** Sample 4	WERF Project#00-HHE-5 Biosolids Odor Quality Test Site 1 June 1, 2002 Location **D** **Headspace** Sample 5	WERF Project#00-HHE-5 Biosolids Odor Quality Test Site 1 June 1, 2002 Location **D** **Headspace** Sample 6

Figure B-2. Example Preprinted Labels.

APPENDIX C: INTERNAL QUALITY CONTROL PLAN

C.1 Quality Assurance Objectives

Quality Assurance Objectives are presented by sample matrix for all sampling and analytical parameters in Tables C-1 and C-2. These values are estimates of the degree of uncertainty that is considered acceptable in order for the data to fulfill the needs of the program. The QA/QC program focuses on controlling and quantifying measurement error within these limits, and provides a basis for understanding the uncertainty associated with these data. In the first step of data validation, measurement data are compared to the QA objectives to determine whether gross performance problems occurred.

Data Quality Objectives (DQOs) are qualitative and quantitative statements that specify the quality of the data to satisfy the end uses of the data to be collected. As such, different data uses may require different levels of data quality. There are five analytical levels that address various data uses and the methods required to achieve the desired level of quality. These levels are divided up as follows:

Screening (DQO Level 1): This provides the lowest data quality but the most rapid results. It is often used for health and safety monitoring at the site, preliminary comparison to local regulations or criteria, initial site characterization to locate areas for subsequent and more accurate analyses, and for engineering screening of alternatives. These types of data include those generated on-site through the use of real-time monitoring equipment at the site like the OVA.

Field Analyses (DQO Level 2): This provides rapid results and better quality than in Level 1. This level may include mobile lab generated data depending on the level of quality control exercised.

Engineering (DQO Level 3): This provides an intermediate level of data quality and is used for site characterization. Engineering analyses may include mobile lab generated data and some analytical lab methods (e.g., laboratory data with quick turnaround used for screening but without full quality control documentation).

Conformational (DQO Level 4): This provides the highest level of data quality and is used for purposes of risk assessment, evaluation of remedial alternatives and principal responsible party (PRP) determination. These analyses require full Contract Laboratory Program (CLP) analytical and data validation procedures in accordance with EPA recognized protocol.

Non-Standard (DQO Level 5): This refers to analyses by non-standard protocols, for example, when exacting detection limits or analysis of an unusual chemical compound is required. These analyses often require method development or adaptation. The level of quality control is usually similar to DQO Level 4 data.

For the most part, laboratory tests will be performed at DQO Level 3, or higher. However, the Standard Tests and olfactometry analysis will be conducted at DQO Level 4.

Table C-1. Data Quality Objectives Matrix
WERF Project 00-HHE-5 Phase II Odor Study.

Parameter	Method	Instrument	Laboratory	DQO Level
Gas Phase Analysis				
Hydrogen Sulfide	Field Instrument	Jerome 631-X Analyzer	Field	2
Ammonia	Field Instrument	Colorimetric Tubes	Field	2
Headspace Reduced Sulfur	GC/MS	HP 5890 GC	VPI	3 or higher
Headspace VFA	GC/MS	HP 5890 GC	VPI	3 or higher
Headspace Amines	GC/MS	HP 5890 GC	VPI	3 or higher
Residual Biological Activity, Methane	GC/Thermal Conductivity detector	HP 5890 GC	VPI	3 or higher
Olfactometry	ASTM E-679		St. Croix	4
Liquids/Solids Phase Analysis				
Soluble Protein	Hartree (1972) modification of Lowry et al. Method (1951)	UV/Visible Spectrophotometer	Bucknell	3 or higher
Labile Protein	Saline Buffer/Shear Extraction followed by Hartree (1972) modification of Lowry et al. Method (1951)	UV/Visible Spectrophotometer	Bucknell	3 or higher
Total Protein	Caustic Extraction followed by Hartree (1972) modification of Lowry et al. Method (1951)	UV/Visible Spectrophotometer	Bucknell	3 or higher
Enzyme Activity	Teuber and Brodisch (1977)	UV/Visible Spectrophotometer	Bucknell	3 or higher
Amino Acid	To be determined	HPLC	Bucknell	3 or higher
Cations- Total	Nitric Acid Digestion (Method 3030E in Standard Methods and analysis of the digest by ICP according to the Method 3500-Fe C)	ICP	Bucknell	3 or higher
Cations- Soluble	Ion Chromatography	Dionex IC	Bucknell	3 or higher
Anions- Sulfate	Ion Chromatography	Dionex IC	Bucknell	3 or higher
Anions- Sulfide	Iodometric Method in Standard Methods	Burette and Standards	Bucknell	3 or higher
Liquid VFA	GC Analysis	GC/FID	Bucknell	3 or higher

Table C-1. Data Quality Objectives Matrix
WERF Project 00-HHE-5 Phase II Odor Study. (cont.)

Parameter	Method	Instrument	Laboratory	DQO Level
RBA, Liquid, Ammonium Ion	Ion Chromatography	DIONEX DX300 IC	VPI	3 or higher
RBA, Liquid, VS Destruction	Standard Methods, 1999	Muffle Furnace and Oven with Scale	VPI	3 or higher

Table C-2. Data Quality Objectives Matrix
WERF Project 00-HHE-5 Phase II Odor Study.

Parameter	Method	Accuracy	Precision	Sensitivity
Gas Phase Analysis				
Hydrogen Sulfide	Jerome 631-X	+/- 20%	+/- 10%	1 ppbv
Hydrogen Sulfide	Colorimetric Tube	+/- 30%	+/- 20%	1 ppmv
Ammonia	Colorimetric Tube	+/- 30%	+/- 20%	1 ppmv
Headspace Reduced Sulfur	GC	+/- 10%	+/- 10%	100 ppbv or lower
Headspace Volatile Fatty Acids	GC	+/- 20%	+/- 20%	100 ppbv or lower
Headspace Amines	GC	+/- 20%	+/- 20%	100 ppbv or lower
Residual Biological Activity, Methane	GC	+/- 5%	+/- 5%	100 ppmv
Odor Concentration	Olfactometry	+/- 50%	+/- 50%	1 DT (Blank Corrected)
Liquids/Solids Phase Analysis				
Soluble Proteins	Hartree (1972) modification of Lowry et al. Method (1951)	+/- 15%	+/- 15%	1 mg/L
Labile Proteins	Hartree (1972) modification of Lowry et al. Method (1951)	+/- 30%	+/- 30%	10 mg/L
Total Proteins	Hartree (1972) modification of Lowry et al. Method (1951)	+/- 30%	+/- 30%	10 mg/L
Enzyme Activity	Teuber and Brodisch (1977)	TBD	TBD	TBD
Amino Acid	TBD	TBD	TBD	TBD
Cations- Iron and Aluminum, Ca, Mg, Na, K –Total	Nitric Acid Digestion (Method 3030E in Standard Methods and analysis of the digest by ICP according to the Method 3500-Fe C)	+/- 15%	+/- 15%	≈50 ppb
Cations–Monovalent and Divalent–Soluble	Ion Chromatography	+/- 10%	+/- 10%	1 mg/L
Anions–Sulfate	Ion Chromatography	+/- 10%	+/- 10%	1 mg/L

Table C-2. Data Quality Objectives Matrix
WERF Project 00-HHE-5 Phase II Odor Study. (cont.)

Parameter	Method	Accuracy	Precision	Sensitivity
Anions–Sulfide	Iodometric Method in Standard Methods	+/- 15%	+/- 15%	0.5 mg/L
Liquid Volatile Fatty Acids	GC analysis	+/- 10%	+/- 10%	TBD
Residual Biological Activity, Liquid, Ammonium Ion	Ion chromatography	+/- 5%	+/- 5%	TBD
Residual Biological Activity, Liquid, Volatile Solids Destruction	Standard Methods, 1999	+/- 5%	+/- 5%	TBD
TBD = To Be Determined				

C.2 Quality Assurance Assessments

The basis for assessing precision, accuracy, completeness, representativeness, and comparability of the data is presented in the following sections.

C.2.1 Precision

Precision measures the reproducibility of repetitive measurements. It is strictly defined as the degree of mutual agreement among independent measurements as to the result of repeated application of the same process under similar conditions. Analytical precision is a measurement of the variability associated with duplicate (two) or replicate (more than two) analyses of the same sample in the laboratory. Total precision is a measurement of the variability associated with the entire sampling and analysis process.

It is determined by analysis of duplicate or replicate field samples, and incorporates the variability caused by matrix variability, field sampling procedures, and analytical variability. The results of total and analytical precision must be interpreted by taking into consideration all possible sources of variability. Duplicate samples will be analyzed to assess field and laboratory precision, and the results will be reported as the relative percent difference (RPD) between duplicate measurements. Analytical precision objectives are presented for each method and matrix in Table C-2.

C.2.2 Accuracy

Accuracy is a statistical measurement of correctness, and includes components of random error (variability due to imprecision) and systematic error (bias). As such, it reflects the total error associated with a measurement. A measurement is accurate when the value reported does not differ from the true value, or known concentration, of the spike or standard. Analytical accuracy is typically measured by determining the percent recovery of known target analytes that are spiked at known concentrations into a field sample. The stated accuracy limits typically apply to spiking levels at five times the method detection limits or higher. The individual methods provide equations for acceptance criteria at lower spiking levels.

Surrogate compound recovery is also reported and is used to assess method performance for each sample analyzed for volatile compounds. Sampling accuracy is assessed by evaluating results for field and trip blanks.

Both accuracy and precision are calculated for specific sampling or analytical batches, and the associated sample results must be interpreted considering these specific measures. An additional consideration in applying accuracy and precision is the concentration level of the samples; a procedure capable of producing the same value within 50% would be considered precise for low-level (near the detection limit) analyses of minor constituents, but would be unacceptable, and possibly useless, for major constituents at high concentrations.

C.2.3 Completeness

Completeness, also referred to as percent data capture, is defined as the percentage of valid data reported compared to the total number of samples collected for analysis. Valid data are determined during the data assessment process and satisfy the QA objectives. Completeness is determined after precision and accuracy are calculated. The objective for completeness for all measurement parameters and all sample matrices is 90%.

C.2.4 Representativeness

Objectives for representativeness will be defined for each sampling and analysis task and will be a function of the investigative objectives. Representativeness will be achieved through use of the standard sampling and analytical procedures described in this General Testing Protocol.

C.2.5 Comparability

Comparability is the confidence with which one data set can be compared to another. The objectives for this QA/QC program are to produce data with the greatest degree of comparability possible. The number of matrices sampled and the range of field conditions encountered must be considered in ultimately determining comparability. Comparability will be achieved by using the same (standard) methods for sampling and analysis, reporting data in standard units, and using standard and comprehensive reporting formats. Analysis of reference samples may also be used to provide additional information that can be used to assess comparability of analytical data produced within the program.

C.3 Laboratory Standards and Reagents

All standards and laboratory reagents, with the exception of common laboratory solvents, are to be dated upon receipt. The preparation and use of all standards are recorded in bound laboratory notebooks that document standard traceability to U.S. EPA or NBS standards. Additional information recorded will include the date of preparation, concentration, name of the preparer, and expiration date, if applicable.

C.4 Method Detection Limit Determination

Following are the steps that will be taken to calculate the method detection limit (MDL) for each analysis:

- Prepare a standard matrix sample at one to five times the estimated MDL (based on the Target Reporting Limit (TRL) and the instrumental detection limit).
- Process seven aliquots of the sample through the entire method.
- Calculate the standard deviation from results of the seven aliquots. The MDL should be equal to the standard deviation times the student's t-value (3.143) for that number of measurements. The MDL shall be equal to or less than the TRL.
- MDLs shall be verified 1 month prior to conducting the analyses. Frequency of this verification shall be stated in the Laboratory's Quality Control Program. If the laboratory has verified an MDL based on the appropriate matrix within these time frames, it does not have to repeat the verification process. All data related to determination and verification of MDLs shall be maintained at the laboratory.
- All field analytical measurement data shall be reduced according to the General Testing Protocol and protocols in applicable SOPs that describe field measurements. Computer programs used for data reduction shall be validated by introducing a test set of data into the program and then comparing the end result of the test set to independently calculated results before use. This verifies the program's operations on a regular basis. Information used in the calculations shall be recorded in sufficient detail to enable reconstruction of the final result at a later date.

C.5 Analytical Data Reduction, Validation, and Reporting

The data reduction, validation, and reporting procedures described in this section will ensure that complete documentation is maintained throughout the program, that transcription and data reduction errors are minimized, that the quality of the data is reviewed and documented, and that the reported results are properly qualified and in a conventional format.

C.5.1 Data Reduction

The reduction of raw data generated at the laboratory bench is the responsibility of the analyst producing it. The data interpretation that is required to calculate sample concentrations follows the methodology described in the specific analytical SOP. After all analyses have been completed, a preliminary laboratory report is generated for review by the laboratory supervisors who verify that the analyses were properly performed and interpreted. After the final review by the laboratory supervisor, the raw data is transferred to sample control and presented for review by the QA coordinator. Raw data, together with all supporting documentation, are stored permanently in confidential files by sample control.

The QA coordinator reviews the data for adherence to the QC method limits. In addition, the data are reviewed for the presence of outliers An outlier is an unusually large (or small) value in a set of observations. There are many possible reasons for outliers, among which are the following:

- Faulty instruments or component parts
- Inaccurate reading of a record, dialing error, etc.
- Errors in transcribing data
- Calculation errors

Sometimes analysts or operators can identify outliers by noting the above types of occurrences when they record observations. In these instances, the errors are corrected, or if correction is not possible the suspect observations may be removed from the data before calculations are performed. If no such information exists, the Dixon Criteria are used to test suspected outliers at the 5% significance level if there are three or more points in the data set containing the outlier. Outliers identified by this method may be removed from the data before further processing (W.J. Dixon, 1953).

A laboratory database is used to store and transfer analytical data from the laboratory. Sample control staff is responsible for entering into the system and verifying sample and result information and generating hard copies of the analytical results.

C.5.2 Data Validation

The designated data validation team will review field documentation and all measurement data for acceptable sample collection and analysis procedures, consistency with expected results or other results, adherence to prescribed QA procedures, and agreement with the acceptance criteria described in Section C.1.

Initially, the reviewer will determine whether hold times were met and that all required analytical QC checks were reported with the data. Then, all QC sample results will be reviewed to evaluate the sampling and analytical performance. Method blank results will be evaluated to identify any systematic contamination; surrogate and duplicate results will be compared to the QA objectives presented in Section C.1, and the results will be used to calculate precision and accuracy for the data set.

This process will identify any analytical methods and compounds for which the QA objectives are not satisfied, and corresponding sample data will be qualified with a flag indicating the problem. Samples collected on the same day, analyzed in the same run or batch, or individual samples may be flagged, depending on the type of problem that has been identified. Reanalysis or resampling may be recommended at this time if data are determined to be unacceptable for the intended application.

The qualifier codes, or "flags," will be stored with the data and printed with the data when reported or transferred for any purpose. The specific statistical procedures and qualifier codes used in the validation process are described in detail in Section 13. After data are received from the laboratory, entered, checked, and qualified, they are a permanent part of the database and cannot be deleted or altered.

C.5.3 Data Reporting

Data reporting for this project will consist of QA reporting, investigative data reporting, and QC data reporting. General reporting practices for measurement data will include the following:

- Heading information identifying the sample batch and the analytical method;
- Unique sample identification number or code;
- Consistent units of measure;
- Consistent number of significant figures;
- No blank or dashed places reported; all spaces will contain a designation (i.e., not analyzed, not sampled, etc.);
- Explanation of outlier values or the cause for deviation from historical data;
- Comparison with regulatory threshold values if applicable;
- Quality assurance flags;
- Quantification of accuracy and precision for analytical data.

C.5.3.1 General Reporting Procedures

The procedures employed to ensure report quality involve the following:

All calculations and measurements will be verified by recalculation by the person initially providing data. The calculations and measurements are then checked by another individual who signs and dates the calculation sheets. Any calculations and measurements that differ from the initial totals are resolved by both individuals. Once the calculations and measurements are included in an internal working copy of a document, the figures are rechecked during peer review. If there are many such calculations within a report, a certain percentage (10%–50%) is checked again during peer review.

Numerical values presented in reports and comparisons of numbers appearing in text, tables, and appendixes will be addressed in the manner discussed above.

C.5.3.2 Investigative Data Reporting

Measurement data generated during the course of an investigation will be reported in tabular form from the computerized database. The formats of the reports will vary, depending on the objectives of the investigation. In general, data will be presented according to Sample Location, analytical method, parameter, and/or matrix. All data will be reported with the qualifiers discussed above, and units will be specified. Commonly used reporting formats will be catalogued and used repeatedly, while specialized formats will be developed as needed. Compound concentration will be reported in ug/m3, and ppbv.

C.5.3.3 QC Data Reporting

Quality control results will be reported by sample matrix and method in tabular form. How these QC results influence the measurement data will be delineated. For example, matrix spike interference will influence specific samples, while laboratory blank contamination will influence all samples extracted or analyzed on a specific day or during a specific analytical run. Two levels of tables may be constructed for each type of QC check. The first level table will contain all QC data, and will present one line per parameter or analysis. First level table formats will be used in presenting duplicate samples and analyses, matrix and method spikes, and system blank results. First level QC data tables will be generated for all investigations.

Specially developed table formats may be used occasionally as an aid to interpretation of the investigative data. The particular format will depend on how the QC results are expected to influence the investigative data. This type of table might be used to identify corresponding investigative results (samples analyzed on corresponding dates) which may be inaccurate. Specialty tables will be generated automatically or manually, depending on the volume of data to be processed and the complexity of the calculations.

C.5.3.4 Field Sampling Report

The field sampling report shall include the following items:

- Narrative of Sampling Event
- The sample log
- Copies of the Chain-of-Custody Sheets
- Laboratory Reports

C.6 Internal Quality Control

Internal Quality Control will be accomplished by collecting and analyzing a series of duplicate, replicate, blank, and matrix spike samples to ensure that the analytical results are within QC limits specified for the program. Laboratory QC samples are documented at the bench and reported with the analytical results. The QC sample results are used to quantify precision and accuracy, and identify any problems or limitations in the associated sample results. Field QC samples will be documented in field logbooks. These components of the sampling program will help produce data of known quality throughout the sampling and analysis component of the program.

C.6.1 Laboratory Quality Control Samples

Laboratory QC is necessary to control the analytical process, to assess the accuracy and precision of analytical results, and to identify assignable causes for atypical analytical results. The QC checks in the laboratory protocol are specific to the analytical method and include the use of one or more of the following QC samples:

C.6.1.1 Method Blanks

A method (or reagent) blank is a sample composed of all the reagents (in the same quantities) used in preparing a real sample for analysis. It is carried through the same sample preparation procedure as a real sample. Method blanks are used to ensure that interferences from the analytical system, reagents, and glassware are under control. The required frequency for analyzing reagent blanks is specified in the analytical SOP for each method, and generally consists of one per day for each method/instrument and/or one per extraction batch.

C.6.1.2 Linearity Check of Calibration Standards

Initial calibration is performed as required for each analytical method, usually using a range of calibration standards with the low standard near the detection limit for the compound. These standards are used to determine the linear dynamic range for the initial instrument calibration, and are usually determined on a monthly basis.

C.6.1.3 Control Samples

Quality control check samples (control samples) are standard samples containing the analytes of interest at a specified concentration, usually in the mid-calibration range. These samples are prepared independent of the calibration standard, and are used to demonstrate that the instrument is operating within acceptable accuracy and precision limits. Quality control check samples are required for GC/MS (off-site) analyses and their preparation and the required frequency of analysis is described in the analytical SOP. They are usually analyzed at the beginning, after every 10 samples, or at the end of an analytical run.

C.6.1.4 Method Spikes

A method spike is a sample of target analytes at known concentrations that is spiked into a field sample before sample preparation and analysis or into the analytical system. Two aliquots of the sample will be spiked and used for the duplicate analysis. The results of the analysis of the duplicate spiked samples are used to measure the percent recovery of each spiked compound and to compare the recovery between samples, which provides an estimate of the accuracy and precision of the method (QA objectives for accuracy are presented in Section C.1). The frequency for method spike analysis is 5-10% of samples analyzed for each method where spikes are performed. Method spikes are sometimes performed in duplicate rather than using field samples in order to obtain precision data for each target compound.

C.6.1.5 Duplicate Analysis

Laboratory duplicates are repeated but independent determinations of the same sample by the same analyst, at essentially the same time and under the same conditions. The sample is split in the laboratory and each fraction is carried through all stages of sample preparation and analysis. Duplicate analyses measure the precision of each analytical method (the method of calculation for precision is outlined in Section C.2.1). Laboratory duplicate analyses are performed for 5% of samples analyzed, or at least one per day, for analytical methods that do not require matrix spike-matrix spike duplicates.

Table C-3 summarizes the specific internal QC checks performed as required for the analytical methods for the tests to be performed in this research. The table also includes information relating to the initial calibration and ongoing calibration checks.

Table C-3. Summary of Laboratory Quality Control WERF Project 00-HHE-5 Phase II Odor Study.

Procedure	QC Check Frequency	Acceptance Criteria	Corrective Action
Headspace Analysis (Organic Reduced Sulfur/TMA/VFA)			
Method Blank	Daily, prior to sample analysis	< 200 ppm or smaller	1-Repeat 2-Clean System and Leak Check
Linearity Check	Monthly	Cor. Coef. > 0.995	1-Repeat Calibration 2-Repeat Linearity Check
Single Point Response Factor (RF) Check	Daily	+/- 20% RPD	1-Repeat RF Check 2-Repeat Calibration
Retention Time (RT) Check	Daily	Agree with RT and MS-spectrum	1-Adjust Instrument 2-Repeat Check
Control Sample	Daily, prior to sample analysis	Correct identification +/- 20% of value	1-Repeat Control Sample 2-Repeat RF Check 3-Repeat RT Check 4-Repeat Calibration
Method Spike			1-Repeat Matrix Sample 2-Repeat RF Check 3-Repeat RT Check 4-Flag Data
Duplicate Analysis		+/- 20% RPD	1-Repeat Analysis
Olfactometry (ASTM E-679)			
Method Blank	> 5%	All	1-Average Blanks and Baseline Subtract
Duplicate Analysis	> 5%	+/- 50% RPD	1-Qualify Data
Organic Analysis (Liquid VFA –BU, C-2 through C-7 Acids)			
Method Blank	Prior to sample analysis	<0.5 mg/L	1. Repeat 2. Check System 3. Create New Blank
Linearity Check	Monthly	r^2 > 0.99	1. Repeat Calibration 2. Repeat Linearity check
Single Point RF Check	1 per batch	+/- 15%	1. Repeat RF Check 2. Check Standards 3. Repeat Calibration
RT Check	Monthly	agrees with established RTs	1. Repeat Check 2. Adjust Instrument
Control Sample	Daily Prior to Analysis	+/-10%	1. Repeat Control Sample 2. Check Standards 3. Repeat RF Check 4. Repeat Calibration
Method Spike	5%, Minimum 1 per Batch	+/-50% recovery	1. Repeat Matrix Sample 2. Repeat RF Check 3. Repeat RT check 4. Flag Data

Table C-3. Summary of Laboratory Quality Control WERF Project 00-HHE-5 Phase II Odor Study. (cont.)

Procedure	QC Check Frequency	Acceptance Criteria	Corrective Action
Duplicate Analysis	5%, Minimum 1 per Batch	+/- 10% RPD	1. Repeat Analysis
Proteins			
Method Blank	1 per batch	<0.5 mg/L	1. Repeat 2. Check Blank 3. Make Up New Blank and Analyze
Linearity Check	1 per batch	r^2>0.99	1. Repeat Calibration 2. Repeat Linearity Check
Single Point RF Check	NA	NA	
RT Check	NA	NA	
Control Sample for soluble samples	1 per batch	+/- 15%	1. Repeat Control Sample 2. Check Standards 3. Make New Control Sample and Check 4. Repeat Calibration
Method Spike	1 per month	+/- 50%	1. Repeat Control Sample 2. Check Standards
Duplicate Analysis	5% minimum, at least 1 per batch	+/- 15% RPD	1. Repeat Matrix Sample 2. Repeat RF Check 3. Repeat RT Check 4. Flag Data
Cations and Anions Analysis on IC (Ca, Mg, Na, K, SO42-)			
Method Blank	1 prior to sample analysis		1. Repeat 2. Check System 3. Create New Blank
Linearity Check	Monthly	r^2 > 0.99	1. Repeat Calibration 2. Repeat Linearity Check
Single Point RF Check	1 per batch	+/- 15%	1. Repeat RF Check 2. Check Standards 3. Repeat Calibration
RT Check	Monthly	agrees with established RTs	1. Repeat Check 2. Adjust Instrument
Control Sample	Daily Prior to Analysis	+/-15%	1. Repeat Control Sample 2. Check Standards 3. Repeat RF Check 4. Repeat Calibration
Method Spike	5%, Minimum 1 per Batch	+/-50% recovery	1. Repeat Matrix Sample 2. Repeat RF Check 3. Repeat RT Check 4. Flag Data
Duplicate Analysis	5%, Minimum 1 per Batch	+/- 15% RPD	1. Repeat Analysis
Total Cations on ICP (Ca, Mg, Na, K, Fe, Al)			

Table C-3. Summary of Laboratory Quality Control WERF Project 00-HHE-5 Phase II Odor Study. (cont.)

Procedure	QC Check Frequency	Acceptance Criteria	Corrective Action
Method Blank	1 Prior to sample analysis	<20 ppb	1. Repeat 2. Check System 3. Create New Blank
Linearity Check	Monthly	$r^2 > 0.99$	1. Repeat Calibration 2. Repeat Linearity Check
Single Point Response Factor Check	1 per batch	+/- 15%	1. Repeat RF Check 2. Check Standards 3. Repeat Calibration
Retention Time Check	NA	NA	NA
Control Sample	Daily Prior to Analysis	+/-15%	1. Repeat Control Sample 2. Check Standards 3. Repeat RF Check 4. Repeat Calibration
Method Spike	5%, Minimum 1 per Batch	+/-50% recovery	1. Repeat Matrix Sample 2. Repeat RF Check 3. Flag Data
Duplicate Analysis	5%, Minimum 1 per Batch	+/- 15% RPD	1. Repeat Analysis
Anions Analysis (total sulfide)			
Method Blank	1 Prior to sample analysis	<1 ppm	1. Repeat 2. Check Standards 3. Create New Blank
Linearity Check	NA	NA	NA
Single Point RFCheck	NA	NA	NA
RT Check	NA	NA	NA
Control Sample	Daily Prior to Analysis	+/-15%	1. Repeat Control Sample 2. Check Standards 3. Repeat Calibration
Method Spike	5%, Minimum 1 per Batch	+/-50% recovery	1. Repeat Matrix Sample 2. Repeat RF Check 3. Repeat RT Check 4. Flag Data
Duplicate Analysis	5%, Minimum 1 per Batch	+/- 15% RPD	1. Repeat Analysis
Enzyme Analysis (metase aminopeptidase)			
Method Blank	1 per batch		
Linearity Check	NA	NA	NA
Single Point RFCheck	NA	NA	NA
RT Check	NA	NA	NA
Control Sample	NA	NA	NA
Method Spike	NA	NA	NA

Table C-3. Summary of Laboratory Quality Control
WERF Project 00-HHE-5 Phase II Odor Study. (cont.)

Procedure	QC Check Frequency	Acceptance Criteria	Corrective Action
Duplicate Analysis	5%, Minimum 1 per Batch	+/- 15% RPD	Repeat Analysis
Amino Acids (met, cys, trp, his)			
Method Blank	TBD	TBD	TBD
Linearity Check	TBD	TBD	TBD
Single Point RFCheck	TBD	TBD	TBD
RT Check	TBD	TBD	TBD
Control Sample	TBD	TBD	TBD
Method Spike	TBD	TBD	TBD
Duplicate Analysis	TBD	TBD	TBD
Duplicate Analysis	TBD	TBD	TBD

C.6.2 Field Quality Control Samples

Field quality control includes quality control for the field instrument(s) and duplicate and blank sample collection and analysis.

C.6.2.1 Field Duplicate Samples

A field duplicate sample is a second sample collected at the same location with the original sample. Duplicate sample results are used to assess precision, including variability associated with both the laboratory analysis and the sample collection process. Duplicate samples will be collected simultaneously or in immediate succession using identical recovery techniques, and treated in an identical manner during storage, transportation, and analysis.

Recovery and analysis of 5% or at least one duplicate sample per day for each method will be performed. Duplicate samples will be taken for both liquid and gas phase samples. The quantities of liquid-phase duplicates required per Sample Location are depicted in Table 5-1, and requirements for field duplicates for the field odor test (for gaseous ammonia and H2S) are depicted in Table C-4.

C.6.2.2 Field Blanks

Field blanks are samples of purified air that are collected and processed in the field using the same sampling and handling procedures as other samples. Field blanks are used to assess the potential introduction of contaminants to the samples during sample collection in the flux chamber and analysis in the laboratory. The frequency requirements for preparing field blanks will be 5% of the gas-phase samples prepared for laboratory analysis. The requirements for field blanks for the field odor test (for gaseous ammonia and H_2S) are depicted in Table C-4.

Table C-4. Field Odor Tests Quality Control
WERF Project 00-HHE-5 Phase II Odor Study.

Procedure	QC Check Frequency	Acceptance Criteria	Corrective Action
Ammonia Analysis (ASTM 4490)			
Field Duplicate	5%	±50% RPD	Flag Data if Necessary
Field Blank	5%	3x MDL	1. Re-Zero 2. Flag Data if Necessary 3. Repeat Check
Jerome Meter Analysis for H_2S			
Replicate	100%	±50% RPD	Flag Data if Necessary
System Blank	Daily	3x MDL	1. Re-Zero 2. Flag Data if Necessary 3. Repeat Check

C.7 Preventative Maintenance

C.7.1 Field Equipment and Instruments

The field equipment for this project includes real-time gas analyzers and the surface flux chamber/support equipment. Specific preventative maintenance procedures to be followed for field equipment are those recommended by the manufacturer. Critical spare parts such as instrument batteries and disposable or expendable items will be kept on-site to minimize instrument down time. Backup instruments and equipment should be available on-site or within one-day shipment to avoid delays in the field schedule.

C.7.2 Laboratory Instruments

As part of their QA/QC Program, a routine preventative maintenance program will be conducted by the laboratories to minimize the occurrence of instrument failure and other system malfunctions. Designated laboratory personnel will be responsible for performing routine scheduled maintenance, and coordinate with the vendor for the repair of all instruments. All laboratory instruments are maintained in accordance with manufacturer's specifications and the requirements of the specific method employed.

This maintenance will be carried out on a regular, scheduled basis, and will be documented in the laboratory instrument service logbook for each instrument. Emergency repair or scheduled manufacture's maintenance is provided under a repair and maintenance contract with factory representatives. Routine preventative maintenance schedules will be provided with the selected laboratories SOPs.

C.8 Accessing Data Precision, Accuracy, and Completeness

The assessment of measurement data is required to ensure that the QA objectives for the project are met, and that quantitative measures of data quality are provided. A distinction must be made between routine quality control and data assessment that is conducted as a part of laboratory operations, and the project-related data assessment process conducted after the data have been reported. It must be assumed that the planning and monitoring that have gone into the sampling and analysis process have served to control the process as much as possible to produce data of sufficient quality for project needs. After the data have been reported, it is necessary to identify any part of the process that could not be controlled, and to what extent that may affect the quality of the reported data.

The routine QC procedures conducted by the laboratories are established in the analytical SOPs. The laboratory is responsible for following those procedures and operating the analytical systems within statistical control limits. These procedures include proper instrument maintenance, calibration checks, and internal QC sample analyses at the required frequencies (i.e., reagent blanks, matrix spike/matrix spike duplicates, laboratory duplicates).

One of the additional ongoing data assessment processes is to maintain control charts for representative QC sample analyses in order to monitor system performance. This provides verification that the system is in statistical control and indicates when performance problems occur, so the problems can be corrected as soon as possible. When reporting the sample data, the laboratory is required to provide the results of associated QC sample analyses.

Problems occur in spite of all precautions taken in planning and execution of the sampling and analysis task. In these cases, the data assessment conducted by The field team leader after the data have been reported must identify the problem, determine which data are affected, and state how these data may be limited for use in the intended applications.

The discussion of data assessment presented in this section pertains to the project-related assessment of data that have been reported after laboratory analyses have been completed. Data assessment procedures established for the testing include:

- Evaluation of blank results to identify systematic contamination
- Statistical calculations for accuracy and precision using the appropriate QC sample results
- Estimation of completeness in terms of the percent of valid data
- Recommendations for corrective actions such as reanalysis or resampling if data are critically affected
- Assignment of data qualifier flags to the data as necessary to reflect limitations identified by the process
- Some basic statistical calculations used in the data assessment process are presented along with a discussion of specific applications to environmental sample results.

C.8.1 Blank Data Assessment

Reagent or method blank results indicate whether any of the contaminants reported in sample results may be attributed to laboratory sources and, therefore, would not likely present in the sampled medium. The most common laboratory contaminants are methylene chloride, phthalates, acetone, and toluene. These are recognized as being ubiquitous in the laboratory environment and controlling them to within acceptable low levels is part of standard laboratory procedure.

If contamination from these compounds is reported in blank samples, the samples associated with the blank—either the same analytical or extraction batch—may be qualified using a data qualifier (B) to indicate that some or all of these compounds may be from laboratory sources. If the concentrations reported in the samples are similar to the blank concentrations, it is likely that all of the contamination was introduced; this assessment is then made in the report for the sampling task.

C.8.2 Accuracy

As previously defined, accuracy is associated with correctness and is a comparison between a measured value and a known, or 'true,' value. Accuracy is calculated from method spike (spikes of the pure matrix), matrix spike results, or the results of the analysis of the PE samples. Spike results are reported by the laboratory as percent recovery and are compared to the accuracy objectives stated in Section 4. Results that do not satisfy the objectives are assigned a data qualifier flag (A) to indicate uncertainty associated with inaccuracy.

Method spikes are spikes of a reference material into a sample matrix (e.g., canister or cryotrap) in the lab. If recovery is outside the established limits, samples from the same batch may be qualified. If any results appear atypical and could be related, those results may also be qualified. The flagged data will be discussed in the report for the sampling task, and specific limitations such as poor or enhanced recovery for specific compounds will be stated.

The percent recovery of matrix spike samples will be calculated using Equation C-1.

$$\%R = \frac{A - B}{C} X100 \qquad \text{(Equation C-1)}$$

where:

A = The analyte concentration determined experimentally from the spiked sample
B = The background level determined by a separate analysis of the unspiked sample
C = The amount of the spike added.

C.8.3 Precision

Precision is a measure of variability between duplicate or replicate analyses and is calculated for field and laboratory replicates. By definition, field precision incorporates laboratory precision. Precision is calculated as the RPD between duplicate analyses or MS or MSD as appropriate. The calculated RPDs are compared to the objectives stated in Section 4. Results that do not satisfy the objectives are assigned a data qualifier flag indicating uncertainty associated with imprecision (P).

An average RPD may be calculated and reported as a measure of overall analytical precision for compounds with multiple measurements. The specific samples collected or analyzed in duplicate are flagged if they do not satisfy the QA objectives. In addition, associated samples may be flagged to indicate variability due to poor precision.

For poor field duplicate precision, samples collected by the same sampling team, from the same equipment, or on the same day may be affected; close evaluation of those results should indicate the most likely source of variability and the corresponding samples will be qualified as warranted. For poor laboratory precision, samples processed and analyzed in the same batch will be more closely evaluated, and any anomalous results will be qualified.

The field team leader is responsible for ensuring that these codes are assigned to the data as required by the established QC criteria, and that they are reported and understood by project staff using the data for specific applications. He is also responsible for initiating corrective actions for analytical problems identified during the QC data assessment process. These corrective actions range from verifying that the method was in statistical control during the analytical runs, to reanalysis of the sample, to resampling.

The relative percent difference (%RPD) will be calculated for each pair of duplicate analysis using the Equation C-2.

$$\%RPD = \frac{S - D}{\frac{S + D}{2}} X100 \qquad \text{(Equation C-2)}$$

where:

S = First sample value (original or MS value)
D = Second sample value (duplicate or MSD value)

C.8.4 Completeness

Completeness is determined after the QC data have been evaluated and the results applied to the measurement data. In addition to results identified as being outside of the QC limits established for the method, the occurrence of matrix effects, and lost samples, samples that could not be analyzed for any other reason are included in the assessment of completeness. The percentage of valid results is reported as completeness Data completeness will be calculated using Equation C-3.

$$\text{Completeness} = \frac{\text{Valid Data Obtained}}{\text{Total Data Planned}} \text{ X 100} \qquad \text{(Equation C-3)}$$

C.9 Corrective Action

During the course of the testing program, it is the responsibility of the WERF Study Team leader to see that all measurement procedures are followed as specified and that measurement data meet the prescribed acceptance criteria. In the event a problem arises, it is imperative that prompt action be taken to correct it.

Problems that require corrective action will be documented by the field team leader as presented in the field log book. He will initiate the corrective action request in the event that QC results exceed acceptability limits or upon identification of some other problem or potential problem. Corrective action may also be initiated by the laboratory coordinator based upon QC data or audit results. Depending upon the severity of the problem, corrective actions range from use of data qualifier flags, to reanalysis of the sample or samples affected, to resampling and reanalysis.

Review of sampling logs and/or analytical results may indicate problems that invalidate the results or critically influence their use. In these cases, corrective action may be required to ensure that valid data are provided. Corrective actions include: recalibration and reanalysis if the analytical system is shown to be out of statistical control; reanalysis if systematic contamination has occurred; resampling if sampling procedures or sample handling have been improper or caused contamination. The severity of the problem and the importance of affected samples will dictate when one of these actions will be required

General recommendations in any case are to follow good laboratory practice and good management practice for all aspects of the sampling and analysis program. These include development and strict adherence to SOPs for all areas, and the establishment of clear responsibilities and lines of communication within the sampling and analytical staff, as well as between project and laboratory staff.

REFERENCES

ASTM Standard Practice E544-99. 1999. *Referencing Suprathreshold Odor Intensity.* American Society for Testing and Materials. West Conshohocken, PA.

ASTM Standard Practice E679-91. 1991. *Determination of Odor and Taste Thresholds by a Forced-Choice Ascending Concentration Series of Limits.* American Society for Testing and Materials. West Conshohocken, PA.

Bak, F.; Finster, K.; Rothfub, F. 1992. Formation of dimethyl sulfide and methanethiol from methoxylated aromatic compounds and inorganic sulfide by newly isolated anaerobic bacteria. *Arch. Microbiol.* 157: 529-534.

Bonnin, C.: Laborie, A.; Pailard, H. 1990. Odor nuisances created by sludge treatment: problems and solutions. *Water Science and Technology*. 22(12): 65-77.

Dentel, S.K.; Gossett, J.M. 1982. Effect of chemical coagulation on anaerobic digestibility of organic materials. *Water Res.* 16: 707-718.

Devai, I.; DeLaune, R.D. 2000. Emissions of reduced gaseous sulfur compounds from wastewater sludge: redox effects. *Environmental Engineering Science*. 17(1).

Dixon, W.J. 1953. Processing Data for Outliers. *Biometrics*. 9: 74-89.

Drotar, A.; Burton, G.A.; Tavernier, J.E.; Fall, R. 1987. Widespread occurrence of bacterial thiol methyltransferases and the biogenic emission of methylated sulfur gases. *Applied and Env. Microbiol.* 53: 1626-1631.

Erdal, Z. K.; Mendenhall, T.C.; Neely, S.K.; Wagoner, D. L.; Quigley, C. 2003. Implementing Improvements in a North Carolina Residuals Management Program. *Proceedings of Water Env. Federation and AWWA Annual Biosolids and Residuals Conference*. Baltimore, MD.

Evanylo, G.K. 1999. Agricultural Land Application of Biosolids in Virginia: Production and Characteristics of Biosolids. Virginia Polytechnic Institute and State University Website: http://www.ext.vt.edu/pubs/compost.

Forbes, R.H.; Adams, G.; Witherspoon, J.; Hentz, L.; Murthy, S.; Glindemann, G.; Higgins, M.; Card, T.; Hargreaves, J.R.; Erdal, Z.K. 2003. Impacts of in-Plant Operational Parameters on Biosolids Odor Quality: Preliminary Results of WERF Phase 2 Study. *Proceedings of Water Env. Federation and AWWA Annual Biosolids and Residuals Conference*. Baltimore, MD.

Glindemann, D.; Novak, J.T.; Murthy, S.N.; Gerwin, S.C.; Forbes, R.H.; Higgins, M. 2004. Standardized biosolids incubation, headspace odor measurement and odor production-consumption cycles. *WEF/A&WMA Odors and Air Emissions Conference.* Bellevue, Washington. Paper in preparation.

Gosset, J.M.; McCarty, P.L.; Wilson, J.C.; Evans, D.S. 1978. Anaerobic digestion of sludge from chemical treatment. *Water Environ. Res.* 50: 533.

Hartree, E.F. 1972. Determination of protein: A modification of the Lowry method that gives a linear photometric response. *Anal. Biochem.* 48: 422-427.

Hentz, L.H., Jr.; Cassel, A. 2000a. The effects of liquid sludge storage on biosolids odor emissions. Water Environment Federation (WEF) 14th Annual Residuals and Biosolids Management Conference. Alexandria, VA.

Higgins, M.J.; Novak, J.T. 1997a. Characterization of exocellular protein and its role in bioflocculation. *Journal of Environmental Engineering*, 123, 479-485.

Higgins, M. J.; Novak, J.T. 1997b. Dewatering and settling of activated sludges: the case for using cation analysis. *Water Environ. Res.* 69: 225.

Higgins, M.J.; Novak, J.T. 1997c. The effect of cations on the settling and dewatering of activated sludges: Laboratory results. *Water Environment Research.* 69: 215-224.

Higgins, M.J.; Murthy, S.N.; Striebig, B.; Hepner, S.; Yamani, S.; Yarosz, D.P.; Toffey, W. 2002. Factors affecting odor production in Philadelphia Water Department Biosolids. *Proceedings Water Env. Fed. Odors and Toxic Air Emissions 2002*, Albuquerque, NM.

Higgins, M.J.; Murthy, S.N.; Novak, J.T.; Yarosz, D.P.; Glindemann, D.; Toffey, W.E.; Abu-Orf, M.M. 2002. Effect of chemical addition on production of volatile sulfur compounds and odor from anaerobically digested biosolids. *Proceedings of Water Env. Fed. 75th Annual Conference.* Chicago, IL.

Higgins, M.J.; Yarosz, D.P.; Chen, Y.C.; Murthy, S.N.; Maas, N.; Cooney, J.; Glindemann, D. 2003. Mechanisms of volatile sulfur compound and odor production in digested biosolids. *Proceedings of Water Env. Federation and AWWA Annual Biosolids and Residuals Conference.* Baltimore, MD.

Kim, H.; Nochetto, C.; McConnell, L.L. Gas phase analysis of trimethylamine, propionic and butyric acid, and reduced sulfurs using solid phase microextraction.

Lowry, O.H.; Rosebrough, N.J.; Farr, A.L.; Randall, R.J. 1951. Protein measurement with the folin phenol reagent. *J. Biol. Chem.*, 193, 265-275.

Murthy, S.N. 1998. Bioflocculation: Implications for Activated Sludge Properties and Wastewater Treatment. PhD Dissertation. Virginia Polytechnic Institute and State University

Murthy, S.N.; Forbes, B.; Burrowes, P.; Esqueda, T.; Glindemann, D.; Novak, J.; Higgins, M.; Mendenhall, T.; Toffey, W.; Peot, C. 2002. Impact of high shear solids processing on odor production from anaerobically digested biosolids. *Proceedings of Water Env. Fed. 75th Annual Conference.* Chicago, IL.

Murthy, S.N.; Peot, C.; North, J.; Novak, J.; Glindemann, D.; Higgins, M. 2002. Characterization and control of reduced sulfur odors from lime-stabilized and digested biosolids, *Proceedings of Water Env. Federation and AWWA Annual Biosolids and Residuals Conference*. Austin, TX.

Murthy, S.; Higgins, M.; Chen, Y.; Toffey, W.; Golembeski, J. 2003. Influence of solids characteristics and dewatering process on volatile sulfur compound production from anaerobically digested biosolids. *Proceedings of Water Env. Federation and AWWA Annual Biosolids and Residuals Conference*. Baltimore, MD.

Novak, J.T.; Glindmann, D.; Murthy, S.N.; Gerwin, S.; Peot, C. 2002. Mechanisms for generation and control of trimethyl amine and dimethyl disulfide from lime stabilized biosolids. *Proceedings Water Env. Fed. Odors and Toxic Air Emissions 2002*, Albuquerque, NM.

Oho, T.; Saito, T.; Koga, T. 2000. Formation of methyl mercaptan from l-methionine by porphyromonas gingivalis. *Infect. Immun.* 68: 6912-6916.

Persson, S. 1992. Hydrogen sulfide and methyl mercaptan in periodontal pockets. *Oral microbiology and immunology*. 7: 378.

Persson, S.; Edlund, M.G.; Claesson, R.; Carlsson, J. 1990. The formation of hydrogen sulfide and methyl mercaptan by oral bacteria. *Oral Microbiology and Immunology*. 5:195-201.

Pohl, M.; Bock, E.; Rinken, M.; Aydin, M.; Konig, W.A. 1984. Volatile sulfur-compounds produced by methionine degrading bacteria and the relationship to concrete corrosion. *Zeitschrift fur Naturforschung C: Biosciences.* 39 (3/4): 240-243.

Ross, D.; Briggs, T.; Bagley, D.; Rupke, M. 2002. The unusual scent of Toronto biosolids: Investigation of the causes and solutions. *Proceedings of Water Env. Federation and AWWA Annual Biosolids and Residuals Conference*. Austin, TX.

Sobeck, D.C.; Higgins, M.J. 2002. Examination of three theories for mechanisms of cation-induced bioflocculation. *Water Research.* 36: 41-52.

Soda, K. 1968. Microdetermination of D-amino acids and D-amino acid oxidase activity with 3-methyl-2-benzothiazolone hydrazone hydrochloride. *Analytical Biochem.* 25: 228-235.

Standard Methods for the Examination of Water and Wastewater. 1999. 18th ed., Am. Public Health Assoc. Washington, D.C.

U.S. Environmental Protection Agency.1999. *Control of Pathogens and Vector Attraction in Sewage Sludge*. Cincinnati, OH.

Teuber, M.; Brodisch, K.E.U. 1977. Enzymatic activities in activated sludge. *European J. Appl. Microbiol.* 4: 185-194.

Water Environment Federation. 1995. *Odor Control in Wastewater Treatment Plants*, WEF Manual of Practice No. 22. New York, USA.

Water Environment Federation. 1998. *Design of Municipal Wastewater Treatment Plants* 4th Ed. WEF Manual of Practice 8.

Winter, P.; Duckham, S.C. 2000. Analysis of volatile odor compounds in digested sewage sludge and aged sewage sludge cake. *Water Science and Technology.* 41(6): 49-55.

Yoshimura, M.; Nakano, Y.; Yamashita, Y.; Oho, T.; Saito, T.; Koga, T. 2000. Formation of methyl mercaptan from L-methionine by Porphyromonas gingivalis. *Infection and Immunity.* 68: 6912-6916.

WASTEWATER UTILITY

Alabama
Montgomery Water Works & Sanitary Sewer Board

Alaska
Anchorage Water & Wastewater Utility

Arizona
Gila Resources
Glendale, City of, Utilities Department
Mesa, City of
Phoenix Water Services Department
Pima County Wastewater Management

Arkansas
Little Rock Wastewater Utility

California
Calaveras County Water District
Central Contra Costa Sanitary District
Contra Costa Water District
Crestline Sanitation District
Delta Diablo Sanitation District
Dublin San Ramon Services District
East Bay Municipal Utility District
El Dorado Irrigation District
Escondido, City of
Fairfield-Suisun Sewer District
Irvine Ranch Water District
Las Virgenes Municipal Water District
Lodi, City of
Los Angeles, City of
Los Angeles County, Sanitation Districts of
Napa Sanitation District
Orange County Sanitation District
Palo Alto, City of
Riverside, City of
Sacramento Regional County Sanitation District
San Diego Metropolitan Wastewater Department, City of
San Francisco, City & County of
San Jose, City of
Santa Barbara, City of
Santa Rosa, City of
South Bayside System Authority
South Orange County Wastewater Authority
Union Sanitary District
West Valley Sanitation District

Colorado
Boulder, City of
Colorado Springs, City of
Littleton/Englewood Water Pollution Control Plant
Metro Wastewater Reclamation District, Denver

Connecticut
New Haven, City of, WPCA

District of Columbia
District of Columbia Water & Sewer Authority

Florida
Broward, County of
Fort Lauderdale, City of
Gainesville Regional Utilities
JEA
Kissimmee, City of, Department of Water Resources
Miami-Dade Water & Sewer Authority
Orange County Utilities Department
Orlando, City of
Reedy Creek Improvement District
Seminole County Environmental Services
St. Petersburg, City of
Stuart Public Utilities
Tallahassee, City of
Tampa, City of
West Palm Beach, City of

Georgia
Atlanta Department of Watershed Management
Augusta, City of
Clayton County Water Authority
Cobb County Water System
Columbus Water Works
Fulton County
Gwinnett County Department of Public Utilities
Macon Water Authority
Savannah, City of

Hawaii
Honolulu, City and County of

Idaho
Pocatello, City of

Illinois
American Bottoms Wastewater Treatment Plant
Dupage, County of
Greater Peoria Sanitary District
Kankakee River Metropolitan Agency
Metropolitan Water Reclamation District of Greater Chicago
North Shore Sanitary District
Wheaton Sanitary District

Indiana
Fort Wayne, City of

Iowa
Ames, City of
Des Moines Metro Wastewater Reclamation Authority

Kansas
Johnson County Unified Wastewater Districts
Unified Government of Wyandotte County/Kansas City, City of

Kentucky
Louisville & Jefferson County Metropolitan Sewer District

Louisiana
Sewerage & Water Board of New Orleans

Maine
Bangor, City of
Portland Water District

Maryland
Anne Arundel County Bureau of Utility Operations
Howard County Department of Public Works
Washington Suburban Sanitary Commission

Massachusetts
Boston Water & Sewer Commission
Upper Blackstone Water Pollution Abatement District

Michigan
Ann Arbor, City of
Detroit, City of
Grand Rapids, City of
Holland Board of Public Works
Lansing, City of
Owosso Mid-Shiawassee County WWTP
Saginaw, City of
Wayne County Department of Environment
Wyoming, City of

Minnesota
Rochester, City of

Missouri
Independence, City of
Kansas City Missouri Water Services Department
Little Blue Valley Sewer District
Metropolitan St. Louis Sewer District

Nebraska
Lincoln Wastewater System

Nevada
Henderson, City of

New Jersey
Passaic Valley Sewerage Commissioners

New York
New York City Department of Environmental Protection

North Carolina
Cary, Town of
Charlotte/Mecklenburg Utilities
Durham, City of
Metropolitan Sewerage District of Buncombe County
Orange Water & Sewer Authority

Ohio
Akron, City of
Butler County Department of Environmental Services
Columbus, City of
Metropolitan Sewer District of Greater Cincinnati
Northeast Ohio Regional Sewer District
Summit, County of

Oklahoma
Tulsa, City of

Oregon
Clean Water Services
Eugene/Springfield Water Pollution Control
Water Environment Services

Pennsylvania
Philadelphia, City of
University Area Joint Authority, State College

South Carolina
Charleston Commissioners of Public Works
Mount Pleasant Waterworks & Sewer Commission
Spartanburg Sanitary Sewer District

Tennessee
Cleveland, City of
Knoxville Utilities Board
Murfreesboro Water & Sewer Department
Nashville Metro Water Services

Texas
Amarillo, City of
Austin, City of
Dallas Water Utilities
Denton, City of
El Paso Water Utilities
Fort Worth, City of
Gulf Coast Waste Disposal Authority
Houston, City of
San Antonio Water System
Trinity River Authority

Utah
Salt Lake City Corporation

Virginia
Alexandria Sanitation Authority
Arlington, County of
Fairfax County Virginia
Hampton Roads Sanitation District
Henrico, County of
Hopewell Regional Wastewater Treatment Facility
Loudoun County Sanitation Authority
Lynchburg Regional WWTP
Prince William County Service Authority
Richmond, City of

Washington
Edmonds, City of
Everett, City of
King County Department of Natural Resources
Seattle Public Utilities
Yakima, City of

Wisconsin
Green Bay Metro Sewerage District
Madison Metropolitan Sewerage District
Milwaukee Metropolitan Sewerage District
Racine, City of
Sheboygan Regional Wastewater Treatment
Wausau Water Works

Australia
South Australian Water Corp.
Sydney Water Corp.
Water Corp. of Western Australia

Canada
Greater Vancouver Regional District
Toronto, City of, Ontario

Mexico
Servicios de Agua y Drenaje de Monterrey, I.P.D.

United Kingdom
Yorkshire Water Services Limited

STORMWATER UTLITY

California
Los Angeles, City of, Department of Public Works
Monterey, City of
Santa Rosa, City of

Colorado
Boulder, City of

Georgia
Griffin, City of

Kansas
Overland Park, City of

Kentucky
Louisville & Jefferson County Metropolitan Sewer District

Maine
Portland Water District

North Carolina
Cary, Town of
Charlotte, City of, Stormwater Services

Pennsylvania
Philadelphia, City of

STATE

Fresno Metropolitan Flood Control District, Calif.
Urban Drainage & Flood Control District, Colo.

CORPORATE

ADS Environmental Services
The ADVENT Group Inc.
Alan Plummer & Associates
Alden Research Laboratory
Alpine Technology Inc.
Aquateam–Norwegian Water Technology Centre A/S
BaySaver Inc.
BioVir Inc.
Black & Veatch
Boyle Engineering Corporation
BPR CSO
Brown & Caldwell
Burns & McDonnell
CABE Associates Inc.
The Cadmus Group
Camp Dresser & McKee Inc.
Carollo Engineers Inc.
Carpenter Environmental Associates Inc.
CDS Technologies Inc.
Chemtrac Systems Inc.
CH2M HILL
Clancy Environmental Consultants Inc.
Damon S. Williams Associates, LLC
David L. Sheridan, P.C.
Earth Tech Inc.
Ecolab Water Care Services
EMA Inc.
The Eshelman Company Inc.
Finkbeiner, Pettis, & Strout Inc.
Frontier Geosciences Inc.
ftn Associates Inc.
Gannett Fleming Inc.
GE Betz
Golder Associates Inc.
Greeley and Hansen LLC
The HACH Company
Hazen & Sawyer, P.C.
HDR Engineering Inc.
HNTB Corporation
HydroQual Inc.
Insituform Technologies Inc.
Institute for Environmental Technology & Industry, Korea
Jacobson Helgoth Consultants Inc.
Jason Consultants Inc.
Jordan, Jones, & Goulding Inc.
KCI Technologies Inc.
Kelly & Weaver, P.C.
Kennedy/Jenks Consultants
Komline Sanderson Engineering Corporation
Lawler, Matusky & Skelly Engineers, LLP
Limno-Tech Inc.
Lombardo Associates Inc.
Malcolm Pirnie Inc.
McKim & Creed
MEC Analytical Systems Inc.
MWH
New England Organics
Odor & Corrosion Technology Consultants Inc. (OCTC)
ONDEO Degremont Inc.
Oswald Green, LLC
PA Government Services Inc.
Parametrix Inc.
Parsons
Post, Buckley, Schuh & Jernigan
R&D Engineering/Conestoga Rover & Associates
The RETEC Group
R.M. Towill Corporation
Ross & Associates Ltd.
Royce Technologies
SAIC Maritime Technical Group
Sear-Brown
Sierra Environmental Services
SYNAGRO
Tetra Tech Inc.
Trojan Technologies Inc.
URS Corporation
USFilter NATC
Vortechnics Inc.
Wade-Trim Inc.
Weston Solutions Inc.
Woodard & Curran
WRc/D&B, LLC
WWETCO, LLC

INDUSTRY

American Electric Power
ChevronTexaco Energy Research & Technology Company
The Coca-Cola Company
Dow Chemical Company
DuPont Company
Eastman Kodak Company
Merck & Company Inc.
ONDEO Services
Procter & Gamble Company
PSEG Services Corp.
Severn Trent Services Inc.
Shell Global Solutions
Thames Water Plc
United Water Services LLC

Note: List as of 5/12/03

Board of Directors

Research Council

Stormwater Technical Advisory Committee

Water Environment
Research Foundation

WERF Product Order Form

As a benefit of joining the Water Environment Research Foundation, subscribers are entitled to receive one complimentary copy of all final reports and other products.Additional copies are available at cost (usually $10).To order your complimentary copy of a report, please write "free" in the unit price column.WERF keeps track of all orders.If the charge differs from what is shown here, we will call to confirm the total before processing.

Name ______________________ Title ______________________

Organization ______________________

Address ______________________

City ____________ State ____________ Zip Code ____________ Country ____________

Phone ____________ Fax ____________ Email ____________

Stock #	Product	Quantity	Unit Price	Total
			Postage & Handling	
			VA Residents Add 4.5% Sales Tax	
			Canadian Residents Add 7% GST	
			TOTAL	

Method of Payment:

(All orders must be prepaid.)

❑ Check or Money Order Enclosed

❑ Visa ❑ Mastercard ❑ American Express

Account No. ____________ Exp. Date ____________

Signature ____________

Shipping & Handling:

Amount of Order	United States	Canada & Mexico	All Others
Up to but not more than:	Add:	Add:	Add:
$20.00	$5.00*	$8.00	50% of amount
30.00	5.50	8.00	40% of amount
40.00	6.00	8.00	
50.00	6.50	14.00	
60.00	7.00	14.00	
80.00	8.00	14.00	
100.00	10.00	21.00	
150.00	12.50	28.00	
200.00	15.00	35.00	
More than $200.00	Add 20% of order	Add 20% of order	

*minimum amount for all orders

Note: Please make checks payable to the Water Environment Research Foundation.

To Order (Subscribers Only):

Log on to www.werf.org and click on the "Product Catalog."

Phone: (703) 684-2470
Fax: (703) 299-0742.

WERF
Attn: Subscriber Services
635 Slaters Lane
Alexandria, VA 22314-1177

To Order (Non-Subscribers):

Non-subscribers may be able to order WERF publications either through WEF (www.wef.org) or IWAP (www.iwapublishing.com).Visit WERF's website at www.werf.org for details.

IWA Publishing's authorised EU representative for General Product
Safety Regulations is Diane D'Arras, 15 rue Duret, 75116 Paris,
France, e-mail: safety@iwap.co.uk.

Printed and bound by CPI Group (UK) Ltd, Croydon, CR0 4YY
12/05/2026
02108868-0018